Environmental Microbiology

Alan H. Varnam

BSc, PhD
Consultant Microbiologist
Southern Biological and
Visiting Lecturer
University of North London, UK

Malcolm G. Evans

Photographer

MANSON
PUBLISHING

A CIP catalogue record for this book is available from the British Library.

For full details of all Manson Publishing Ltd titles please write to:
Manson Publishing Ltd
73 Corringham Road
London NW11 7DL, UK.

Project management, design and layout: Paul Bennett
Text editing: Michael Meakin
Colour reproduction: Tenon & Polert Colour Scanning Ltd, Hong Kong
Printed by: Grafos SA, Barcelona, Spain

CONTENTS

Preface   4

Glossary   5

Acknowledgements   6

1 Overview   7
The Nature of Microbial Communities   8
Life at Low Nutrient Concentrations: the Common Lot of Micro-organisms   16
Physicochemical Factors Affecting the Environmental Fate of Micro-organisms   18
Competitive Strategies of Micro-organisms   23
Interactions Involving Micro-organisms   25
References   27

2 Aquatic Environments   29
Microflora of the Different Aquatic Environments   30
The Marine Environment   33
The Freshwater Environment   56
References   84

3 Terrestrial Environments   87
Soil   88
Micro-organisms and Higher Plants   105
Other Terrestrial Habitats   122
References   128

4 Extreme Environments   131
High Temperature Environments   132
Extremely Acidic Environments   137
Highly Alkaline Environments   140
Non-alkaline, Highly Saline Environments   144
References   151

Appendices   152
I Development of the Microflora in a Flooded Gravel Pit   152
II The Great Goldmine Mystery   153
III A Comparison of the Microbiological Profile of the Soil on Two
   Sides of a Valley   154

Further Reading   155

Index   156

PREFACE

Adam had 'em
The world's shortest poem: on the antiquity of microbes

Micro-organisms are ubiquitous. They are present in the soil of our gardens, in the air that we breathe and in the sea around us. They invade and colonize our bodies, both causing illness and being protective of health. They spoil our foods and yet are involved in the manufacture of some. Micro-organisms are able to grow in a remarkable range of habitats, which range from the relatively hospitable, often supporting a wide diversity of microbial life, to the extremely hostile, where only a single type may proliferate. An example of an extremely hostile environment, which has attracted much recent interest, is the vent of undersea volcanoes – the realm of the hyper-thermophile. Despite enormous differences in the nature and geobiology of the various environments, it should be appreciated that the same underlying factors govern the life and death of micro-organisms irrespective of the nature of the environment. Equally, their behaviour is governed by fundamental rules.

A vast amount of work, both at bench and theoretical level has been necessary over a period of many years to enable us to assemble the body of knowledge concerning environmental microbiology available today. Until relatively recently, bench work has been primarily based on conventional cultural techniques and, providing the inherent limitations are understood, these can still be of considerable value. In recent years, however, an array of powerful new tools, many based on genetic or advanced microscopic techniques, have become available. Advances at bench level have been complemented by the development of new concepts at the theoretical level.

The study of environmental micro-organisms is not merely an academic pursuit. Environmental micro-organisms have long been exploited, knowingly or unknowingly, in the management of Planet Earth and in the economic activities of humans. Our increased understanding of the interactions between micro-organisms and the environment have enabled a refinement of their exploitation. An example is the bioremediation of soil polluted by oil during the Gulf War.

This book is intended to provide a sound, fundamental knowledge of environmental micro-biology, the principles being reinforced by a wide range of examples, illustrated in colour where appropriate. The interactions between the environment and the dominant microflora is discussed as are the interactions between the various components of the microflora. The emphasis, however, is on understanding principles. This book is neither intended to be a long list of every micro-organism which may be present in a given environment, nor is it intended to provide detailed descriptions of the taxonomy of the micro-organisms. Such information, if required, may be obtained elsewhere.

As environmental microbiology is a difficult subject area to divide into well defined and rigid categories, the usual chapter-based format has not been used. The book is divided into four main parts: an Overview of environmental microbiology, Aquatic, Terrestrial and Extreme environments, each subdivided into major subject areas. It is intended that this rather unconventional approach will enhance the ease of use. Above all, however, it is hoped that the book will go at least some way to show what a challenging and fascinating topic environmental microbiology is, even when experimental results appear to defy interpretation.

Alan H. Varnam
Malcolm G. Evans

GLOSSARY

Definitions are given in the context of micro-organisms, although similar terms may be applied to other forms of life.

Acidophile. A micro-organism which grows optimally at pH values below 5.0. Should not be confused with acid-tolerant micro-organisms. A **thermoacidophile** grows both at high temperatures and low pH value.

Actinomycete. A diverse group of Gram-positive eubacteria, amongst which a mycelial growth habit is common.

Alkaliphile. A micro-organism which grows optimally at pH values above 8.0.

Allochthonous. Micro-organisms which are not consistently a major part of the microflora of an environment, but which become of significance in response to an increase in nutrient concentrations, or other favourable event. Known previously as **zymogenous**.

Archaebacteria. A primary subgroup of cellular life, comprising bacteria of prokaryotic cell structure, but a highly distinctive cell chemistry, which differs profoundly from that of eubacteria. Also known collectively as **archae** and individually as an **archeon**.

Autochthonous. Micro-organisms which are consistently present in significant numbers in an environment and form the indigenous microflora.

Autotroph. An organism capable of growth using only inorganic nutrients. Typified by green plants, autotrophic micro-organisms include **photoautotrophs** (light as their energy source and CO_2 as their C source) and **chemoautotrophs** (chemicals as their energy source). Also known as **chemolithotrophs**.

Bloom. Seasonal dense growth of algae or other planktonic micro-organisms on water.

Copiotroph. A micro-organism which requires a high level of nutrients for growth.

Coryneform. Gram-positive bacteria, of characteristic irregular morphology, similar to that of *Corynebacterium*. These are numerically important in a wide range of habitats, including soil.

Cyanobacteria. Photosynthetic eubacteria, previously known as 'blue-green' algae.

Denitrification. Anaerobic bacterial respiration of nitrate, resulting in reduction to nitrite and, by a cascade of two further anaerobic respirations, to N_2 gas. As part of the nitrogen cycle, it is important in geomicrobiology, but also occurs in nitrite-containing (cured) foods.

Epiphyte. A micro-organism which grows on the surface of plants, but takes no water or nutrients from the plant.

Eubacteria. A primary subgroup of cellular life, comprising most bacteria. Cells are prokaryotic in structure, but the cell chemistry is similar to that of eukaryotes.

Eukaryote. An organism in which the unit of structure is the complex eukaryotic cell. Eukarotes are of extraordinary diversity, including all known life forms except the prokaryotes.

Eutrophication. The process of enrichment of a body of water (usually freshwater) with inorganic nutrients, such as nitrate and phosphate. **Eutrophic** is the term applied to a body of water, which has been enriched in this way. Eutrophication can be a natural process, but is often due to the works of man.

Hadal. The ocean deep below 11,000 m.

Halophile. A micro-organism which has an obligate requirement for NaCl at greater than physiological levels. May be sub-classified as moderate and extreme. The term should not be applied to micro-organisms which are NaCl tolerant, but which have no requirement. **Haloalkilophiles** are halophilic micro-organisms which require a high pH for growth.

Heterotroph. An organism, exemplified by animals, which requires organic nutrients.

Humus. The organic fraction of soil, consisting of naturally occurring compounds resistant to degradation by micro-organisms.

Lichen. A composite organism consisting of a fungus living in association with one, or sometimes two, symbionts. Symbionts may either be algae or cyanobacteria.

Lithobiotic. Living in association with rocks and stones. **Endoliths** live in the interior of stones and **epiliths** on the surface.

Methanogens. One of the constituent groups of the archaebacteria. Anaerobic micro-organisms which convert fermentation products of other anaerobes to methane.

Methophils. A group of eubacteria having the

ability to use single carbon compounds, including methane, as its sole source of carbon and energy. Previously known as methylotrophs.

Mineralization. The decomposition of plant and animal remains and excretion products to inorganic material, which may be utilized by photosynthetic organisms.

Mycorrhiza. Composite structure formed by higher plant roots in symbiotic relationship with a fungus.

Myxobacteria. The distinct group of eubacteria having a gliding motility and, in many cases, a special developmental cycle which involves formation of a fruiting body.

Neuston. The layer at the surface of bodies of water.

Nitrate respiration. The use of nitrate in place of oxygen as the terminal electron acceptor in the growth of some aerobic eubacteria.

Nitrification. The geochemical process in which specialized eubacteria oxidize ammonia to nitrite and nitrite to nitrate. An important part of the nitrogen cycle.

Oligotroph. A micro-organism capable of growth at low nutrient levels.

Parasitism. A symbiotic relationship in which benefit is obtained by one partner only, the parasite. The second partner often suffers damage.

Pelagic. Relating to the open sea. The term is sub-classified as **epipelagic** (surface waters above thermocline), **mesopelagic** (waters directly below thermocline), **bathypelagic** (waters at depth *ca.* 2,000 m) and **abyssopelagic** (water at depth below 3,800 m).

Phototroph. A photosynthetic organism, either prokaryotic or eukaryotic.

Plankton. Floating micro-organisms in aquatic habitats. The term is not restricted to unicellular algae, planktonic bacteria may be described as **bacterioplankton**, ultramicrobacteria as **picoplankton** and bacteriophage as **femtoplankton**. **Zooplankton** describes free floating micro-organisms classified with the animal kingdom. **Phytoplankton** describes all photosynthetic planktonic organisms.

Prokaryote. Organisms in which the unit of structure is the relatively simple prokaryotic cell. These are members of two microbial groups, the eubacteria and the archaebacteria.

Prosthecate. Bacteria characterized by the formation of cell extensions (prosthecae). These are filiform or conical, the interior being continuous with the cytoplasm of the main part of the cell.

Rhizosphere. Regions of soil immediately surrounding plant roots together with the root surfaces of that plant.

Saprophyte. An organism which obtains nutrients directly from dead, or decaying, organic matter.

Symbiosis A strategy for meeting biological competition by adapting to life in close and continuing association with another form of life. Partners are known as symbionts, or bionts, e.g. a mycobiont being a mould partner in a symbiosis and a photobiont, a photosynthetic partner.

Stratification. Formation, in bodies of water, of horizontal layers of distinct chemical and/or physical properties. The boundary is known as the thermocline or chemocline, and stratification may either be temporary (**holomitic**) or permanent (**meromitic**).

Sulphate-reducing bacteria. Strictly anaerobic eubacteria, capable of anaerobic respiration utilizing oxidized S compounds as the electron acceptor.

Acknowledgements

The authors gratefully acknowledge the help of many persons during the writing and preparation of this book. Particular thanks are due to:

Debbie and Phil Andrews for producing many of the hand-drawn and computer-generated drawings and diagrams.

David Post of Oxoid Ltd, for the gift of selective media.

Nikon (UK) Ltd for the loan of a photographic microscope.

The libraries of the BBSRC Institute of Food Research, Reading Laboratory and the University of Reading for their help in obtaining information.

Our colleagues, both in the UK and overseas, who have permitted us to reproduce illustrations derived from their research. *Individual acknowledgments accompany the relevant caption.*

Colleagues, both in Reading and elsewhere, for their help and interest during the preparation of the book.

1 OVERVIEW

Until the 1960s many aspects of environmental microbiology were dealt with on an observational basis. Certainly there was a good measure of understanding of economically important phenomena, such as the symbiotic relationship between *Rhizobium* and leguminous plants, while there was a basic understanding of the relationship between microbial growth and physical parameters, such as temperature and of the various types of interaction between different micro-organisms in the same environment. Overall, however, many observations remained poorly explained and, to many microbiologists, it appeared that the various paradigms used to explain and interpret the behaviour of micro-organisms in natural environments were deficient. Over the past two decades, the situation has changed and while, happily for environmental microbiologists, much remains less than fully understood, there now exists a framework of knowledge which has greatly expanded the understanding of the relationship between micro-organisms and their environments. Two examples are an understanding of biofilm formation and its environmental importance and of the survival strategies of non-endospore forming micro-organisms.

THE NATURE OF MICROBIAL COMMUNITIES

Introduction

Natural environments are usually inhabited by a diverse population of micro-organisms. These can encompass a wide range of physiological and nutritional types, from autotrophs to heterotrophs, from psychrophiles to hyperthermophiles and from obligate aerobes through micro-aerophilic bacteria to oxygen-sensitive anaerobes. There are exceptions to this general rule, usually where the environment has been manipulated by man or where the extreme physicochemical nature limits the types able to grow. Examples of the latter include highly alkaline and saline lakes and highly acidic hot springs. Even under these extreme conditions, however, diversity can be much greater than was thought possible in earlier years – in recent years many other concepts in environmental microbiology have been replaced. It was previously considered, for example, that micro-organisms dwelt in discrete compartments for each physiological and nutritional type but it is now recognized that most microbial ecosystems are heterogeneous. More recently still, advances in methodology, in the framework of challenges to earlier orthodoxies, have gone a considerable way to elucidate the development and function of such diverse populations.

Spatial organization

It is now accepted that the indigenous microflora of most environments develop in spatially organized physicochemical gradients (Keith *et al.*, 1987); the distance over which these gradients are generated varies very considerably. Microbial films (biofilms) may be only a few micrometres thick in low-nutrient environments, while soil crumbs, microbial mats, sewage flocs and biofilms in nutrient-rich environments are of macroscopic rather than microscopic dimensions. Less commonly, gradients may extend over many metres. Such situations, which include stratified lakes, sewage outfalls and geothermal springs, can be seen as exceptional but important environments.

It is the existence of a physicochemical gradient that permits the development and coexistence of a heterogeneous population of micro-organisms. The microbial population is organized either horizontally or vertically depending on the direction of the gradient. In many environments a wide variety of physiological and nutritional types are present and micro-organisms with apparently conflicting lifestyles can grow in close proximity. This is a particularly marked phenomenon in physically small environments where the gradients can be very steep. The oxygen gradient across small soil particles, for example, permits the growth of aerobes, micro-aerophils and anaerobes within a physically very short distance.

Heterogeneous communities are of major importance in the microbial world because of the considerable advantages gained by members of the population. Indeed it has been stated that spatial organization in biofilms and similar situations permits micro-organisms to obtain many of the benefits of multicellular life. Interaction between micro-organisms permits activities such as co-metabolism and cross-feeding, while diverse populations are less affected by environmental change and can recover from disaster more rapidly than ecosystems of lower diversity. This leads to long-term stability, although it should be appreciated that most microbial ecosystems are dynamic communities and are subject to continual short-term changes. Changes can result from either periodic or non-periodic events affecting either the physicochemistry of the environment as a whole, or the gradients within a given environment. The physicochemical effects are both direct (through the immediate effect on a given part of the population) and indirect (through the effect on interactions between members of the community). It may be considered, however, that short-term changes enhance long-term stability.

Biofilms

A number of definitions of biofilm have been proposed. The most satisfactory is considered to be that 'a biofilm is a community of microbes embedded in an organic polymer matrix, adhering to a surface' (*Feature 1*).

Feature 1. Biofilms and man

In all known habitats, bacteria preferentially reproduce on a surface rather than in suspension in the liquid phase. Biofilms were first extensively studied during the 1940s but it was not until the 1970s that it was appreciated that their formation occurs in almost all natural environments. 'A rock immersed in a stream, an implant in the human body, a tooth, a water pipe or conduit, etc. are all sites where biofilms develop' (Carpentier and Cerf, 1993). Given the extent to which biofilms impinge on man and his works, the belated recognition of their full importance is rather surprising. From the medical viewpoint, colonization of implants can serve as focal points for potentially fatal infections, the cells in deep layers being protected from antibodies and phagocytic white blood cells (Anwar and Costerton, 1992). Biofilm formation is also a key factor in the pathology of dental disease. In the food industry, it is thought that development of biofilms on machinery surfaces serves to protect pathogens such as *Listeria monocytogenes* from sanitizing agents, while in other industries biofilms are responsible for substantial economic loss (*Table 1*). The world of microbiology is never simple of course and biofilm formation can be exploited by man. *Acetobacter* spp., growing in a biofilm for example, are responsible for the oxidation of ethanol to acetic acid in traditional vinegar manufacture. More commonly, in modern industrial applications, exploitation of biofilms involves their high efficiency in entrapping molecules and small particulate matter and the development of consortia able to degrade even highly recalcitrant molecules (*Table 2*).

Table 1. Examples of industrial problems caused by biofilm formation.

Industry	Problem
Fluid transfer (general)	Reduction of flow rates and blocking of pipes
Potable water distribution	Reduction of flow rates
	Taints
	Unacceptably high bacterial numbers
Paper making	Poor quality paper
Flour milling; malting; sugar refining	Interference with operation through slime formation
Ships' hulls	Increase in fuel consumption; lower overall speeds
Structural steelwork, pipelines, etc.	Accelerated corrosion
Heating and cooling operations	Reduced efficiency
Food processing	Reservoir of spoilage and potentially pathogenic micro-organisms; possible survival of pathogens through under-processing
Printing	Blockage of ink supply pipes; reduced ink quality

Table 2. Processes involving the exploitation of biofilm formation, with some examples.

Process	Example of exploitation
Water purification	Slow sand filtration
Sewage purification	Percolating filters
Agricultural waste purification	Floccor towers
Industrial waste purification	Bioreactors
Removal of heavy metal pollution (including uranium)	Mercury waste remediation
Recovery of metals from low-grade ores	Microbial leaching
Production of acetic acid from ethanol	Traditional process

A number of physiological changes occur in micro-organisms immediately after adhesion, these changes increasing in magnitude with time. Most biofilms in natural habitats consist of an assemblage of different micro-organisms and are heterogeneous in both structure and chemical composition. In all cases, however, the major constituent of the matrix is water. The remainder consists of various extracellular polymers including glycoproteins and polysaccharides, the nature of which is poorly understood. These are usually referred to as extracellular polymeric substances (EPS) and comprise 50–95% of the dry weight of the biofilm. In the past, it was thought that micro-organisms were more or less evenly distributed throughout the biofilm. This may be the case in some circumstances but the usual structure is a collection of micro-colonies, separated by 'water channels' (1, 2).

Biofilms confer on their inhabitants the benefits of a stable environment enjoyed by members of most heterogeneous microbial communities. Biofilms also serve to physically protect microbial cells from external factors such as ultraviolet (UV) irradiation and heat, and to minimize the effects of changes in physical parameters such as pH as well as offering protection from inhibitory substances. Protection from inhibitors can be particularly important where microbial growth is controlled by biocides and disinfectants but it is also of considerable significance in permitting the survival of bacteria following severe pollution.

In addition, the entrapment of water by biofilms minimizes the risk of dehydration. This is of particular significance to micro-organisms in aquatic

1 Microcolonies within a biofilm developing on the surfaces of a glass flow cell. Images obtained in cross-section show microcolonies forming individual mounds of cells.

2 Microcolonies within a biofilm developing on the surfaces of a glass flow cell. Images obtained through the film show microcolonies forming individual mounds of cells. **1** and **2** reproduced with permission from Woolfardt G.M., *et al.*, 1994, © 1994 The American Society for Microbiology.

3 Diagrammatic representation of biofilm formation. Initially both bacteria (red) and particulate material (green) are suspended freely above the surface to be colonized.

4 Following **3** there is a concentration to the surface.

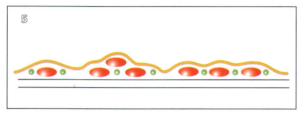

5 Following **4** there is the formation of the 'young' biofilm.

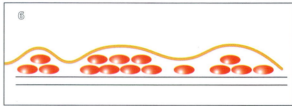

6 As cell multiplication continues, the biofilm thickens to the point at which 'sloughing off' occurs.

environments, periodically exposed to drying-out, as well as in soils, which are subject to seasonal dehydration. As well as protection from physical and chemical stress, biofilms shield bacteria from phages (Hicks and Rowbury, 1987). It may also be that there is protection from lytic and parasitic bacteria as well as from eukaryotic predators. At the same time, predatory micro-organisms may well benefit from the high population of bacteria in the vicinity of biofilms as a result of sloughing off of the outer layers and the release of daughter cells.

Nutrient diffusion through biofilms tends to be slow and only limited quantities are available to micro-organisms in the deep layers. Metabolic activity by such cells is limited and their status has been described as 'quiescent' (Lewis and Gattie, 1990). This status does have the advantage that the bacteria are not subject to competition. At the same time, biofilms are highly effective in trapping nutrients in the form of organic and mineral molecules or discrete particles. This results from the overall negative charge on the polysaccharide molecules within the biofilm matrix.

The formation of biofilms follows a defined sequence, differing on chemically inert surfaces and in living tissue. The fundamental difference lies in the fact that only non-specific adhesion is involved in the environment, while specific adhesion, by means of lectins and adhesins, is a basic process in the development of biofilms in living tissue though only the development of biofilms in the natural environment will be discussed in detail here.

Biofilm formation on chemically inert surfaces involves four stages: transport, adsorption of molecules, adhesion of micro-organisms and colonization (3–6). Transport involves the movement of molecules, inert particles and micro-organisms to the surface on which the biofilm subsequently forms and involves several processes. This is followed by adsorption of molecules, which occurs almost instantaneously. The accumulation of molecules at the solid–liquid interface may be referred to as the 'conditioning film'. Microbial growth rates are enhanced by the higher nutrient concentration at the interface. Adhesion occurs much more slowly than adsorption and involves two distinct stages. In the first, reversible, stage the bacteria continue to show Brownian movement and are easily removed by rinsing or other simple mechanical procedures. The second stage is regarded as irreversible and involves much stronger binding. The degree of adhesion of individual types of micro-organism is determined by the surface properties which, in turn, are determined by the EPS. There is also variation according to the phase of growth of the cell. Exceptions are known, but

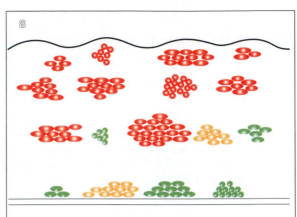

7 Diagrammatic representation of nutrient limitation in biofilms. Even in a relatively newly established biofilm some cells in the lower layer become 'quiescent' (coloured yellow in diagram).

8 As the thickness in **7** increases some colonies in the lower layers become moribund (coloured green in the diagram) while an increasing number move to the 'quiescent' state.

in the majority of cases, EPS released at the end of the exponential growth phase, or during the stationary (starvation) phase, favours adhesion to hydrophobic surfaces. This is consistent with the finding that, in a marine environment, submerged surfaces are usually first colonized by small cells, a small cell size usually being associated with starvation (Roszak and Colwell, 1987). There are obvious advantages for starving bacteria in being primary colonizers in the area of relatively high nutrient concentration at the solid–liquid interface. Extrinsic factors, such as growth temperature may, however, also be involved. Colonization is the final stage and leads directly to formation of the biofilm. Organisms producing EPS multiply within the developing matrix, ultimately forming microcolonies. The rate of colonization varies considerably and the time taken to reach an equilibrium state ranges from a few hours to several months.

A biofilm is a dynamic rather than a static entity, which inevitably undergoes changes with time. The outer surface is continually changing as cells both peel away and adhere. Cells which peel away are usually the more hydrophilic daughter cells. Biofilms can, therefore, be seen to provide a continuing inoculum of daughter cells. This is of environmental importance in that cells, adapted to a given environment, are available to initiate biofilm formation on new surfaces, such as a branch fallen into a stream, or on existing surfaces which, for some reason, have not previously been colonized.

Despite sloughing off, biofilms tend to increase in thickness, leading to changes in the physicochemical gradient. Events within the biofilm can be affected by many factors, including changes in nutrient and O_2 availability, and pH. In the case of nutrients, cells close to the surface of the biofilm receive sufficient to support reproduction, in contrast to the situation in deeper layers where nutrient levels may only be sufficient for maintenance, or even lower. As the thickness of the biofilm increases, conditions within the sub-surface layers change (**7, 8**). The most obvious effect is that the availability of nutrients is decreased leading to previously active cells becoming quiescent. This effect is not absolute and cells of bacteria of lower nutrient requirements, or which are more readily adaptable to alternative substrates, will continue to reproduce for longer periods, thus changing the balance between different species. At some point, however, the thickness of the film becomes such that the nutrient supply in the deepest layers is insufficient to supply even maintenance requirements of the quiescent bacteria. Cell death ensues, leading to the film becoming fragile and sloughing off from the substrate. Biofilm thickness increases most rapidly during periods of high nutrient availability which are favourable to the actively growing cells in the outer layers of the biofilm. Film stability may, however, be reduced by the increasing thickness leading rapidly to the point at which nutrient diffusion to the deeper layers effectively ceases and the death of the microorganisms in those layers.

The effect of stress and disaster on microbial communities

All microbial communities are subject to stress as a result of changes in environmental conditions significantly outside normal parameters. On some occasions stress may be sufficiently severe to cause a disastrous (destructive) effect on the community involving the death of all, or a high proportion, of its population. Stress may result from natural changes or from the works of man. Many of the more dramatic changes are attributed to man and may be deliberate (e.g. sterilization of food) or accidental (e.g. release of pollutants with strong anti-microbial activity). Dramatic changes can also be caused by natural events such as wildfires.

Stress has been defined in various ways, but a practical definition is 'an abiotic factor or set of factors which limit the production of biomass'.

Stress factors are typically unfavourable temperature, pH value, nutrient supply, water availability, and the like. Stress does not, of course, affect all members of a population equally and as a general rule normal parameters are much narrower for eukaryotic organisms than prokaryotic.

The effect of stress obviously depends on the nature, extent and rapidity with which change occurred. Caution is required, however, in that small changes in environmental conditions can have much greater impact than predicted. Where change is slow, members of microbial populations can, within limits, adapt (habituate) to changing environmental parameters and changes over protracted periods may have relatively limited effects (*Feature 2*).

Feature 2. Storm and stress

Concerns over undesirable and possibly dangerous environmental changes are, with justification, a preoccupation of the 1990s. We tend, however, to think in terms of direct threats to man and the macro-environment. Examples are the connection between the diminution of the ozone layer and skin cancer, and deforestation due to acid rain. It is less commonly recognized, even by environmentalists, that these and other factors also place considerable stress on microbial communities. Such stress may affect man indirectly, through loss of productivity and consequent effects on the food chain to higher life, or directly through loss of desirable properties exploited in activities such as bioremediation. Micro-organisms do, however, have a remarkable capacity to adapt and this may be illustrated by a brief examination of a potentially disastrous situation: acidification.

Acidification of water and soils in some parts of the world has led to serious problems. Amongst the best publicized is the damage to forest trees caused by acid rain or snow, at least partly a consequence of the burning of high-sulphur coal in power stations. Acidification may also result from agricultural practices and industrial pollution. The slow development of acidification means that a high level of induced tolerance (habituation) is possible. This permits micro-organisms to survive in acidic environments which would rapidly be lethal to their non-habituated counterparts. Acid tolerance may also induce cross-tolerance to other types of stress. It is thought that phosphate may have a role in enhancing ability for survival and growth at sub-optimal pH values. This is of obvious consequence in high-phosphate environments.

Adaptation has often been considered a property restricted to Gram-negative bacteria but it has also been demonstrated for a few Gram-positive genera, including *Listeria monocytogenes*. Even amongst the Gram-negative genera there is considerable variation. A very high degree of adaptation is possible for *Citrobacter*, for example, but very little for *Serratia*.

Although acidification is generally considered to arise from human activity, natural causes, such as the emission of sulphur dioxide during volcanic eruptions, can also be involved. Archaeological studies in the north of Scotland in 1996 produced evidence of a stable neolithic community destroyed as a result of crop failure. This is thought to have been due to acid rain following a major volcanic eruption. Data concerning adaptation is based on Nojoumi *et al.*, 1995.

Despite adaptation, stress is often marked by a decrease in diversity. This may be seen in the eukaryotic population, but not in the prokaryotic, but in extreme cases the population is reduced to a single dominant species. This is seen within foods where processing is manipulated to reduce the number of potential spoilage organisms. Equally, however, highly stressful environments may be exploited by specialized micro-organisms (such as thermoacidophiles) as a means of minimizing competition. In such situations microbial productivity can be very high in the absence of mutual competition between members of a diverse community and consequent reduction of growth rates. This phenomenon may also be seen when apparently favourable changes result indirectly in stress through increased competition. Increased availability of nutrients, for example, which would enhance growth of most members of a community in pure culture, can lead to a reduction in diversity during aquatic blooms as a consequence of competition stress.

In natural environments recovery from disasters often involves re-colonization from unaffected areas. Immediately after the disaster, however, surviving organisms, often endospore-forming bacteria, are able to dominate the microflora. The microflora usually remains in a non-equilibrium state during initial stages of re-colonization. In many cases, cyanobacteria are the primary colonizers of denuded areas. Ultimately, however, a stable community will be re-established. The relationship of this community to that existing before largely depends on the extent to which the physicochemical nature of the environment has changed. Even where the physicochemical conditions return rapidly to those pertaining before the disaster, however, re-establishment of the microflora can be a very slow process. Under some conditions, the reasons for this are obscure.

Definition of the structure and diversity of the constituent population

Defining the structure and diversity of natural microbial communities through quantification of the constituent population has long been seen as one of the major objectives in environmental microbiology. Conventional approaches invariably involve culture followed by some form of classification, including the identification of isolates according to prevailing taxonomic criteria, determination of physiological or biochemical properties, or simply according to morphology, etc. Such approaches suffer from severe limitations in that conventional culture techniques have been estimated to recover only between 1 and 10% of those micro-organisms present, although some microbiologists believe that such poor recoveries are a consequence of inadequate technique. It has, for example, been possible to culture the dominant bacteria in seasonal blooms in near-shore waters (Rehnstam *et al.*, 1993). Many problems appear to stem from the continuing use of rich culture media which select for copiotrophic micro-organisms rather than the oligotrophs present in many natural environments. Despite these difficulties, cultural methods are still of use in many situations providing that the inherent limitations are fully understood.

Definitive techniques based on non-cultural methods are, however, enabling the understanding of the environment to be taken beyond the boundaries imposed by the limitations of traditional techniques. Microscopy has undergone something of a renaissance, with the application of new techniques enabling sophisticated visualization of the morphology of micro-organisms in their environments to be made. Assay for characteristic cellular constituents has also been a widely used approach. Fatty acid analysis is one of the most successful of these techniques, allowing for the estimation of biomass and, by use of biomarkers, determination of the incidence of specific groups of micro-organisms (Rajendram *et al.*, 1994). The greatest overall impact, however, has been made by genetic techniques.

Genetic techniques are, of course, well established. In the early days, when the use of guanosine plus cytosine (G+C) ratios was providing a platform to underpin phenotypic taxonomy, genetic techniques appeared to be solely in the realm of the taxonomists. Later, genetic techniques were applied to detection of specific micro-organisms, usually pathogens, in medical, and subsequently, food microbiology. DNA hybridization was widely used, being designed either to detect a whole organism or probes to detect genes encoding a specific factor – usually, in medical bacteriology, virulence. An important development was the polymerase chain reaction, in its various forms, which offers a highly sensitive means of detecting both live and dead cells, a feature which can both be an advantage and a disadvantage.

Genetic techniques are widely used in environmental microbiology, the number and diversity of applications possibly being the greatest of any branch of the science. Both probes and the polymerase chain reaction have been used in studies of microbial communities, although currently the most powerful tool in many applications is the determination of 16S rRNA or 16S rRNA F1 encoding DNA (rDNA). A 16S rRNA probe was used to determine evidence for indigenous *Streptomyces* populations in a marine environment, for example, (Moran *et al.*, 1995), but the capabilities of the tool were perhaps demonstrated more fully by Moyer *et al.* (1994). These workers were able to estimate the diversity and community structure of a microbial mat at an active hydrothermal vent system by determination of variations in 16S rRNA genes. This would simply not have been possible without genetic techniques.

It is inevitable that genetic techniques will be further developed. Development is a two-way process: as our knowledge of the environment expands, we require increasingly sophisticated tools to explore further and as those sophisticated tools are developed we see new opportunities for their application. There is only one note of caution: however valuable any technique is, it is not a philosopher's stone. Every technique has its limitations and it is necessary to be aware of, and understand, these if application is to be successful.

LIFE AT LOW NUTRIENT CONCENTRATIONS: THE COMMON LOT OF MICRO-ORGANISMS

The general use of nutritionally rich media for *in vitro* cultivation of micro-organisms can lead to a misunderstanding of the nutrient status of natural environments. In most natural environments, oligotrophic conditions are the rule rather than the exception and a continuing state of starvation is the lot of many micro-organisms. Even in environments such as the animal body, which appear as a 'horn of plenty' for micro-organisms, nutrient deficiencies may limit growth. Iron deprivation, for example, is an important factor in limiting growth of invasive micro-organisms and reduction in plasma iron levels (hypoferraemia of infection) is an equally important defensive response to bacterial invasion. Some foods do represent a nutrient-rich environment but, as in many artificial media, metabolism is profligate and growth of individual bacteria becomes limited both by deficiencies in key nutrients and by build-up of metabolites.

In many cases the majority of bacteria present are heterotrophic (chemo-organotrophic). Heterotrophs in natural environments are usually limited by the availability of carbon; the rate of carbon-associated energy flow ultimately determining multiplication, maintenance or death of cells. At low carbon levels, energy requirements are growth rate dependent, in contrast to the growth rate independent model for high carbon levels (Muller and Babel, 1996). Available energy is always partitioned between growth and maintenance. Maintenance processes have a higher affinity for ATP but cannot command the total energy supply (Chesbro *et al.*, 1979). It should be appreciated, however, that while carbon is the common limiting factor for heterotrophs, low levels of inorganic phosphates may also limit population development (although under some circumstances phosphates are present in abundance). Phosphate limitation is an example of starvation stress and is a particular phenomenon of natural waters, although inorganic phosphate limitation can occur elsewhere. Paradoxically, these environments can contain high levels of organic phosphates derived from biological or anthropomorphic sources, but utilization requires the micro-organism to be able to synthesize phosphatases.

This ability appears to impart a significant competitive advantage (Ozkanka and Flint, 1996).

In environments such as water and soil, heterotrophic micro-organisms may be classified as oligotrophs and copiotrophs. Oligotrophs are characterized by a low growth rate, highly efficient substrate scavenging and accumulation of nutrients, over a long period, to permit multiplication. This lifestyle is the norm even at relatively high nutrient levels. Copiotrophs, at high nutrient levels, are of low substrate affinity and high growth rate. At limiting nutrient levels, however, copiotrophs use a number of survival strategies. Complex molecular regulatory mechanisms are activated in response to starvation. Expression of starvation genes results in a reduction in size (**9**) and the formation of 'ultramicrobacteria', usually by a process of reductive division (**10, 11**). Protective 'starvation proteins' are synthesized and substrate capturing enhanced. Amino acid uptake, for example, switches from low affinity to high affinity in response to nutrient limitation. In at least some cases, such cells are culturable providing sufficient care is taken (**12**). All bacteria, however, have a point at which exogenous nutrients can no longer be obtained and are thus effectively exhausted. At this point cells may enter a dormant, anabiotic state that permits them to survive for long periods without division (Kaprelyants *et al.*, 1993). During this state cells are considered to be nonculturable. It also seems likely that in some situations at least true dormancy is preceded by a 'nutritionally opportunistic' state in which endogenous metabolism is sufficient to maintain the ability of the cell to respond very rapidly to nutrient availability. A low level of endogenous metabolism, together with a low maintenance energy requirement, would therefore appear to be of considerable importance not only in determining the capacity of micro-organisms to survive, but also in responding rapidly to the advent of more favourable conditions. *Aeromonas hydrophila* is an example of a micro-organism that can adapt to a wide range of nutrient concentrations. It appears, however, that at low nutrient levels the advantages of this inherent property are often undone by the presence of autochthonous micro-organisms. This is probably a result of competition for the available nutrients (Kersters *et al.*, 1996).

It should be mentioned that observations have solely involved aerobic bacteria and that the situation with obligate anaerobes may be entirely different.

There has been some interest in using ultra-microbacteria in biotechnology. Major interest has involved the use of genetic modification to provide non-multiplying cells which can express a high level of a desired biochemical activity in dense fermenter cultures. A further possibility is use in bioremediation where conventional methods are limited by low nutrient levels leading to low metabolic activity.

9 Bacterial cells under conditions of nutrient starvation. A marine *Vibrio* underwent a rod-coccus transformation while held in artificial seawater at room temperature for 20 days. About 50% of cells could be recovered by direct plating on to non-selective artificial seawater medium. (Microscopy: phase contrast, × 1,000.)

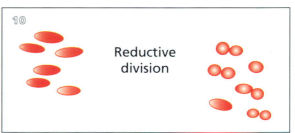

10 At least some bacteria have a viable nonculturable response resulting from a temperature downshift which is distinct from, but related to, the starvation response. This is a strain of *Vibrio vulnificus* that underwent reductive division in seawater at 25°C (77°F) but remained recoverable by standard procedures for a prolonged period. From Oliver *et al.*, 1995.

11 In seawater at 5°C (41°F) cells underwent rod-coccus transformation but reductive division was not involved and the cells rapidly become nonculturable. From Oliver *et al.*, 1995.

12 Entry into the nonculturable state by *V. vulnificus* in seawater at 25 °C (77°F) and 5 °C (41°F). After Oliver *et al.*, 1995.

PHYSICOCHEMICAL FACTORS AFFECTING THE ENVIRONMENTAL FATE OF MICRO-ORGANISMS

As discussed above, the constituents of microbial communities are determined by the physicochemical conditions and by nutrient availability. These may be seen as determining the broad composition of communities in that similar physiological and nutritional types will be selected at each niche. In many cases dominance within each niche will be determined by the competitive ability of each micro-organism against others selected by the same environmental pressures (see pages 23, 24).

It is common to describe the influence of physicochemical factors on micro-organisms in terms of parameters determined in pure culture. This must be treated with extreme caution, especially since the effect of one parameter is usually determined when other parameters are optimal. Equally it is necessary to be aware, when evaluating the effects of parameters such as pH value, that the physicochemical status of the environment as a whole may be very different to that at the level of the micro-organism. Conventional means of measuring parameters are of little value and the use of microelectrodes is required. These have tip diameters in the region of 50–100 µm although diameters of less than 10 µm are common in the determination of dissolved O_2 (Herbert, 1992).

A great many environmental parameters affect the fate of micro-organisms. Those of greatest *overall* importance will be discussed here (temperature, composition of atmosphere, pH value, water availability and light radiation). It is recognized that other factors, including both those naturally occurring (such as concentrations of salts) and 'artificial' factors (such as added preservatives in foods and other industrial products) are important in some instances, some of which are discussed later. At the same time, the present discussion is largely restricted to non-extreme environments; extreme environments are discussed in Extreme Environments (page 131). Some overlap between extreme and non-extreme environments is, however, inevitable.

It is also necessary to appreciate that the relative importance of the different factors varies according to the environment and the individual micro-organism. It is possible to envisage each factor as a hurdle of varying height (the hurdle concept) each of which must be 'jumped' by a micro-organism to be able to develop in an environment. This concept is useful but does not fully take account of interactions between physicochemical factors. These are very difficult to predict, but advantage may be taken of recently developed mathematical modelling techniques.

A mathematical model is a set of equations representing a prototype (a representation of the system under study). Two main types of model have evolved (Bazin and Prosser, 1992), the first – the mechanistic model – is used to explain how an ecosystem operates. The second – the predictive model – is used to forecast future events.

Temperature

Temperature is demonstrably of very considerable importance in determining microbial growth rates and metabolic activity. In many environments, however, low temperatures are common and it is unsurprising that these habitats have been colonized by cold-adapted micro-organisms, able to grow relatively rapidly at temperatures of *c*.0°C (32°F). Such micro-organisms may be considered to be of two types, psychrotrophs (which are able to grow at low temperatures but have a maximum temperature above 20°C (68°F)) and psychrophiles (which have an optimum for growth of 15°C (59°F) or lower and are unable to grow at temperatures above 20°C). In contrast to very high temperatures, temperatures near 0°C are not considered 'extreme' and a wide range of micro-organisms – comprising representatives of both Gram-negative and Gram-positive bacteria, cyanobacteria, fungi and eukaryotic algae – are either psychrotrophic or psychrophilic. Many are members of genera that are common in higher temperature environments. In contrast to the general rule, however, a number of psychrophilic bacteria have been isolated that do not have direct counterparts in warmer environments. These include '*Micrococcus cryophilus*' (which has a typical Gram-negative cell envelope and is not a species of *Micrococcus*), *Arthrobacter glacialis*, *Aquaspirillum arcticum* and 'Vibrio psychroerythrus'.

Psychrotrophic micro-organisms are found in a wide range of habitats (**13**) including those that are permanently cold and those where temperatures fluctuate on either a regular or irregular basis. In contrast, psychrophilic micro-organisms are found only in environments that are permanently below 5°C (41°F). The most important of these is marine waters, 90% of the volume of the ocean being below 5°C. Water in the depths of deep mountain lakes is also permanently cold. Permanently cold soils and sediments are found in the polar regions, in caves and in the vicinity of ice-fields. Even in such permanently cold habitats psychrotrophs outnumber psychrophiles. It has been suggested, however, that in at least some cold habitats psychrophiles represent the autochthonous microflora and psychrotrophs the allochthonous flora (Baross and Morita, 1978).

Life at low temperatures requires significant adaptation to maintain replicative ability. The efficiencies of electron transport, ion pumping and nutrient uptake have been described as determinants of psychrophilly and depend on membrane lipid fluidity and protein activity.

Micro-organisms produce stress proteins in response to a shift to low temperatures. The psychrotrophic yeast *Trichosporon pullulans*, for example, produced 26 stress proteins in response to a 21 to 5°C (70 to 41°F) cold shock but only six proteins in response to a 15 to 5°C (59 to 41°F) shock. In the case of micro-organisms in habitats that are permanently cold, proteins necessary for life at low temperatures must be permanently synthesized as a result of cold adaptation. Such proteins must be distinguished from stress proteins which confer a transient protection (Gounot, 1991). Environmental micro-organisms unable to grow at seasonal low temperatures also have survival mechanisms which appear related to, but distinct from, mechanisms for survival at very low nutrient concentrations. A limited degree of adaptation is possible to low temperature growth although the significance is not clear. This is illustrated by the fact that in cold environments the optimal temperature for growth and carbon metabolism is usually very significantly higher than the *in situ* temperature. At higher temperatures there is a much closer adaptation of growth and metabolism to ambient temperature.

Until recently there has been relatively little interest in the use of psychrotrophic and psychro-

13 The importance of low temperatures in extending the storage life of food was recognized empirically many years before systematic studies of the relationship between temperature and microbial growth. Large-scale production of ice at central plants was common before cheap mechanical refrigeration became available to individual shops and the like.

philic micro-organisms in biotechnology. It is recognized, however, that the ability of such micro-organisms and their enzymes to be metabolically active at low temperatures has potential advantages in a number of areas. It has also been recognized that cold environments, especially those which are stressful in other ways, such as low nutrient availability, high salinity and high pH value, may be an important source of micro-organisms with novel enzymes of potential industrial application.

In pure culture, many micro-organisms are less tolerant of temperatures above their optimum than those below. In many natural environments, the effect of temperature increase is very significant and can lead to stress of the population even when increases are small. In many cases this is probably an indirect effect, resulting from increased competition and other factors. Protection against high temperatures can be conferred by stress proteins and limited adaptation is possible both to growth at higher temperatures and to withstanding temperatures in the lethal range.

arid and select a specialized microflora. A specialist microflora is also stress selected in foods and other materials preserved by drying; this usually comprises moulds. A common environmental situation is periodic lack of water due to climatic conditions. This requires survival rather than growth and micro-organisms produce a variety of desiccation-resistant bodies including both endospores and exospores, many of which can persist for extended periods of time. Vegetative cells often gain protection from biofilms but can be remarkably desiccation resistant without any apparent protection. Coccal-shaped cells are more resistant than rod-shaped and drying-out of an environment can change the balance of the population after rehydration.

Light radiation

Terrestrial and aquatic micro-organisms are exposed to both visible and UV radiation. In some circumstances, such as survival of enteric micro-organisms in seawater, light of both visible and UV wavelengths is a major factor. It has been hypothesized that the underlying cause is oxidative stress (Goumelon *et al.*, 1994). Susceptibility to photo-inactivation depends both on species and individual strains, many pigments having a significant protective role. Photosensitizers may be endogenous within the cell, while exogenous photosensitizers, such as humic substances, can also be involved. Susceptibility, especially to UV radiation, is usually significantly greater in micro-organisms from human and animal sources, where exposure is minimal, than in similar micro-organisms from the external environment. This may be demonstrated by reference to *Leptospira*, the free-living *L. biflexa* being far more resistant than the parasite *L. interrogans*. Resistance to UV radiation is, however, highly conserved in subsurface micro-organisms, which are effectively shielded.

Light is, of course, essential to primary production through photosynthesis. Higher plants are adapted to use the highest intensity of light available, but the exposure of most algae and cyanobacteria to light of full intensity would lead to rapid bleaching, loss of photosynthetic ability and ultimately death. Sensitivity to UV-B (280–315 nm) is greatest and populations can be under stress even at normal levels. Increased UV-B irradiation, for example resulting from the thinning of the stratospheric ozone layer, has decreased primary production by phytoplankton in Antarctica

by 6–12%. Decreases in primary production have been noted elsewhere. A number of strategies have been developed amongst these micro-organisms to avoid harmful effects while gaining greatest benefit from exposure to light. Some, for example, produce 'sunscreen' compounds which absorb UV-B and thus reduce the likelihood of damage to target cellular compounds. An alternative strategy involves motility and a movement to the zone of optimum light intensity, where conditions permit photo-synthesis without reducing growth. An exception is the red freshwater flagellate *Euglena sanguinea* which contains photoprotective pigments and which orientates itself to light (Gerber and Hader, 1994). This enables the flagellate to take advantage of high light intensities and the stress-induced lack of competition at the water surface.

No strategy for minimizing exposure to UV radiation, however, can be entirely successful in preventing a degree of damage. The damage primarily involves DNA, which absorbs UV light, the energy so released causing dimerization between adjacent pyrimidine residues on the same DNA strand. The occurrence of the pyrimidine dimers triggers a repair mechanism which excises them, exposing a short region of single-stranded DNA on the opposite strand. Single-stranded regions are also formed if a replication fork passes a pyrimidine dimer that has not been excised. The formation of the single-stranded regions induces the *SOS system*, a general DNA repair system. This, in turn, induces at least nine enzymes that fill the gaps opposite the single-stranded. The SOS system is rapid acting, but relatively inaccurate, being described as error-prone repair. This accounts for the mutations common after exposure to UV and, of course, the deliberate use of UV as a mutagen in genetic studies. A second repair system involves excising defective sections of a DNA strand and their replacement by insertion of new nucleotides. This system is relatively error free. The relative importance of the two systems varies in different bacteria. Bacteria also differ in overall susceptibility to UV damage. Radiation-resistant bacteria, such as *Deinococcus*, for example, have highly effective DNA repair mechanisms, although it is also possible that the DNA of radiation-resistant bacteria is inherently more stable than that of the less resistant. Research has shown, however, that negative responses to UV irradiation are not necessarily due to exposure to UV-B alone but that interactions with the effects of UV-A (320–400 nm) and 'photosynthetic active radiation' (400–700 nm) are also of major importance (Nielsen and Eklund, 1995).

COMPETITIVE STRATEGIES OF MICRO-ORGANISMS

Competition is often considered to involve micro-organisms that have the same *overall response* to the environmental pressures present and thus share the same niche. Considerable care is needed, however, in using even relatively simple definitions. It is necessary, for example, to emphasize overall response to environmental pressures. It is a common misconception that the outcome of competition between two, or more, micro-organisms is decided by one defining factor, such as pH. Except in the very simplest of environments, however, the most successful organism is that which can respond most effectively to a multi-factorial matrix of stress factors. At the same time, key factors may be unrelated. The outcome of competition between two micro-organisms in a given environment may be determined by whether the benefit gained by one organism by its response to Factor A (e.g. efficiency of nutrient utilization) is greater than the benefit gained by the other organism in response to Factor B (e.g. low pH value).

In many cases, competition is between micro-organisms which, physiologically at least, are closely related. This is fairly obvious, since two micro-organisms which respond in the same, or similar, ways to environmental stress are likely to compete in the same niche. In some cases, the competing micro-organisms are strains of the same species. In wine-making, using a natural yeast inoculant, for example, the succession of different karyotypes of *Saccharomyces cervisiae* results from the relative competitive ability of each in conditions of falling sugar and rising alcohol content. It is important, however, to avoid anthropomorphism. It may justifiably be argued that the existence of karyotypes able to compete effectively in conditions ranging from the relatively hospitable to the relatively hostile is itself a strategy to ensure that *Saccharomyces cerevisiae* is able to compete over a wide range of environmental conditions. A similar overall strategy has been adopted by a number of micro-organisms. Many moderately halophilic bacteria, for example, can grow over a remarkably wide range of NaCl concentrations, being the dominant organism at some concentrations, but secondary at others (see page 146). Other micro-organisms have two, or more, sets of key enzymes that are induced in response to differences in environmental circumstances, such as reduced temperatures. The ability to adapt to unfavourable conditions, either by production of stress proteins, or by other routes, also permits some micro-organisms to compete, whether successfully or not, outside their normal range.

Competition does not always involve physiologically similar micro-organisms, but care is required to decide which situations are truly competitive. The ability of aerobic and anaerobic bacteria to coexist in soil, for example, was long an enigma for soil microbiologists, since it would be expected that one category would rapidly attain dominance. The advent of mini-probes, which measure physico-chemical parameters over very small areas, demonstrated that very steep gradients can exist between adjacent, micro-scale areas. Thus aerobes and anaerobes are effectively capable of growth side-by-side. This cannot be considered competition unless the situation is complicated by a stress factor affecting both aerobes and anaerobes, such as nutrient deprivation.

True competition between physiologically distinct micro-organisms does exist, an example being that in micro-oxic environments between microaerophilic bacteria and some aerobes. In such situations, it would be expected that micro-aerophilic bacteria would be dominant. This is not, however, the case when aerobic bacteria, such as *Pseudomonas*, which can compete effectively in a wide range of environments, are present. Strains of *Pseudomonas* inhabiting micoaerophilic environments, such as those of reed beds (see page 78) have remarkably effective cytochrome systems, which are able to scavenge oxygen imparting a competitive advantage.

In an overall appreciation, three main competitive strategies available to micro-organisms have been identified, although many micro-organisms employ a combination of two, or three. In environments such as most soils and natural waters, which are capable of supporting a large population of micro-organisms, the most common is directly competitive, or combative, and involves competition for nutrients, including greater nutritional versatility. Antagonists specific to other micro-organisms, such as

14 Production of inhibitors by some colonies is a common feature of crowded isolation plates from environments such as soil. It *must not*, however, be assumed that such phenomena on artificial media have any significance in the environment. (Source: garden soil; Medium: nutrient agar; Incubation: 22°C (71.6°F), 4 days).

bacteriocins and antibiotics (**14**) may be produced and there may be non-specific antagonism due to environmental modification by, for example, production of fermentation acids. In the case of fungi, there may also be aggressive mycelial interactions.

The second strategy, ruderal, involves the ability of the minority sub-set of the population to respond more rapidly to improved environmental conditions than the current majority sub-set and subsequently attain dominance. The most common environmental condition involved is nutrient status; but other factors, including temperature, pH value and dilution of NaCl levels due to rainfall, can all be of importance. Micro-organisms effective under conditions of ruderal competition are characterized by their rapid uptake of easily assimilable substrates and fast growth rate. Long-term colonization depends on the period over which the changed conditions persist, but is likely to be difficult. That sub-set of the population which is able to take advantage of an improvement in environmental conditions must have a strategy for survival during less favourable times. It has been postulated that the highly heat- and desiccation-resistant endospores of *Bacillus* and *Clostridium* fulfil that role, while fungi and actinomycetes produce exospores which not

only are desiccation-resistant, but which can be widely disseminated and thus colonize new areas where conditions are favourable.

It must be appreciated, however, that not all micro-organisms involved in ruderal competition have specific survival mechanisms for unfavourable periods. In some cases, the population falls when nutrient levels, or other favourable factors, can no longer sustain rapid growth and a large population. To be able to take advantage of an improvement in conditions it is probable that a minimum population must be maintained. This may well be accompanied by a marked reduction in metabolic rate and the initiation of the various survival strategies. These, for example, may involve downsizing of cells, usually by reductive division (see page 16), and a switch from low affinity substrate uptake to high affinity.

The third strategy is not really competitive at all in that the organism exploits an extremely unfavourable habitat where competition is effectively non-existent. The best example is that of thermoacidophiles (see page 133). A considerable degree of physiological adaptation is required, but long-term colonization is possible, albeit in a very restricted number of niches.

INTERACTIONS INVOLVING MICRO-ORGANISMS

Micro-organisms enter a vast range of symbiotic relationships involving one or more partners, which may be other micro-organisms, animals or plants. Micro-organisms can also be a third party in many plant–animal relationships. It should be noted that various definitions are applied to symbiosis – that used here is broadly based and covers any living association no matter how transient (Evans, 1988).

Various definitions have been applied to interactions involving micro-organisms. In some cases it is difficult to determine the nature of the symbiosis. The ectomycorrhizal symbiosis between fungi and plants (see page 108), for example, is considered by some mycologists to be a directed parasitism of either the plant by the fungus or vice versa (Malloch, 1988). More generally the symbiosis has been described as a co-evolved mutualism (Pirozynski and Hawksworth, 1988).

Competition: Competition (see above) is probably the most common interaction, although the number and complexity of mutualistic interactions (see below) may be much greater than previously thought. Where the outcome of competition depends on response to a single limiting factor, the development of both populations is likely to be restricted. In more complex situations, however, where the outcome of competition depends, for example, on the adequacy of response to two different limiting factors, only one population is likely to be restricted. This may lead to total dominance by one population, although co-existence is possible, even when one population is considerably larger.

Amensalism: Amensalism is the condition where the outcome of competition invariably leads to complete dominance by one population, which restricts the growth of its competitor population(s) without itself being affected. Changing conditions can lead to a reversal of the relationship between the populations. This may be drastic, or gradual. Extensive rainfall on highly saline lakes can lead to a dilution that is catastrophic for the previously dominant extremely halophilic archaebacteria and cause dominance by the previously subordinate moderately halophilic eubacteria (see page 147). This reversal is usually fairly short-term. Gradual reversal of the relationship occurs, for example,

when a lake is subject to slow acidification (see page 76). The reversal is usually long-term, although there may be further changes in the balance of the microflora if acidification continues.

Commensalism: Commensalism involves benefit to one population while the other is unaffected. A quoted example is the production, by *Arthrobacter globiformis*, of the 'terregens' factor, one of a number of sideramines essential to the growth of soil-dwelling 'coryneform bacteria'. Some microbiologists, however, consider this relationship to be actually mutualistic. This underlines the fact that commensalism is difficult to define and many relationships described as commensal are, in fact, mutualistic or parasitic. Commensalistic relationships are also dynamic and may become parasitic, or mutualistic. Equally, parasitic, or mutualistic relationships may become commensalistic.

Mutualism: Mutualism exists in a very large number of forms. In some, there is mutual dependence and neither micro-organism can grow in the absence of the other. Such dependent relationships are usually stable. Mutualism may, equally, involve no more than a casual relationship, which may be transitory. Mutualism is of considerable importance in the development and maintenance of microbial communities, especially when nutrient levels and other limiting factors for growth are subject to change.

A **microbial consortium** is a stable mixed culture, such as can be found in biofilms. The relationships between members of the consortium are usually, but not invariably, mutualistic.

Parasitism: In both parasitism and the parallel process, predation, one organism benefits directly at the expense of the other. In parasitism the smaller organism usually benefits and infection often leads to the death of the larger host cell. In some cases, it is important to distinguish between the effect of parasitic infection on individual cells and on populations as a whole. Parasitic attack by bacteriophage in the aquatic environment recycles a high proportion of nutrients, which may be utilized directly by surviving micro-organisms.

Predation: In predation, the benefiting organism is usually the larger, although there are exceptions;

the predatory bacterium *Bdellovibrio bacteriovorus*, for example, is considerably smaller than its prey. A number of predatory mechanisms exist and many predators locate prey by means of complex 'signals'. Some aquatic bacteria are known to produce a proportion of large cells amongst the general population and it has been postulated that this is a means of defence against 'small' predators, which do not recognize the prey signal. The large cells are, however, more susceptible to predation by 'large' predators.

Indirect interactions: Interactions are often discussed in terms of micro-organisms impinging directly on each other. This is not, however, correct and many indirect interactions exist, which are of considerable environmental significance. An example is the behaviour of micro-organisms during natural fermentation (involving pH fall) of decaying vegetation. The fermentation is usually initiated by environmental members of the Enterobacteriaceae, acid production, together with a rise in the local CO_2 concentration within the vegetation, creating conditions for growth and further acid production by *Lacobacillus* and other lactic acid bacteria. When the fermentable substrate is exhausted, acid-tolerant, lactate-utilizing yeasts become dominant, the pH value rises and a variety of other bacteria develop following exhaustion of lactate and a fall in the yeast population. Successions of this nature are widespread in the natural environment and in foods. In general, the micro-organism furthest down the succession sequence is the most successful. There are, inevitably, exceptions. *Brevibacterium linens*, which is involved in the ripening of many varieties of surface ripened cheese, initiates changes that permit the growth of lactate-tolerant yeast and subsequently *Penicillium*. Under some, ill-defined, circumstances, *Br. linens* remains the dominant micro-organism despite extensive growth of the yeast and mould. (*Note that this is a fault in varieties such as Camembert and Brie.*)

A further example of an indirect interaction is that between microalgae and aerobic bacteria. Many factors may be involved, but an interesting hypothesis is that the bacteria, (*Pseudomonas* spp.) stimulate the algae (*Scenedesmus* and *Chlorella*) by consumption of photosynthetically produced oxygen. The reduction of O_2 levels in the micro-environment consequently produces ideal conditions for photosynthetic growth. In addition to the benefit obtained by the ready availability of O_2, the bacteria benefit from the release of assimilable carbon compounds by the algae.

Interspecies hydrogen transfer. Methanogenic bacteria have highly efficient hydrogenases. The affinity of these enzymes for their substrate (K_m) is sufficiently high to maintain a vanishingly low partial pressure of H_2 when active methanogenesis is occurring. The resulting low P_{H2} creates conditions which allow a number of fermentative anaerobes to re-oxidize NADH by means of an NADH-coupled hydrogenase:

$$NADH + H^+ \rightleftarrows NAD^+ + H2$$

Although hydrogenases of this type are widely distributed among anaerobes, they are unable to re-oxidize NADH when grown in pure culture, because the equilibrium lies too far to the left. When grown in culture with methanogens present, the constant removal of hydrogen by methanogenesis results in a P_{H2} sufficiently low to shift the equilibrium towards the right and allow the reaction to proceed. In turn, the fermentation balance shifts towards more oxidized products with higher yields of ATP and biomass. Interspecies hydrogen transfer may be seen as a synergistic relationship.

Relationships among a wide range of micro-organisms are discussed in later pages. These include the lichen, mycorrhizal and nitrogen-fixing symbioses, the latter two, in particular being of considerable economic as well as environmental importance (see pages 108–111). Although the examples of economically important symbioses discussed in detail concern plant:micro-organism interactions, the importance of animal: micro-organisms should not be forgotten, although considered beyond the scope of this book. These include the role of micro-organisms in the health and illness of man and other animals and the economic importance of the rumen symbiosis in cattle and sheep.

REFERENCES

Anwar, H. and Costerton, J.W. (1992) Effective use of antibiotics in treatment of biofilm-associated infections, *American Society of Microbiology News*, **58**, 665–8.

Baross, J.A. and Morita, R.Y. (1978) Microbial life at low temperatures: ecological aspects, *Microbial Life in Extreme Environments* (ed. D.J. Kushner), pp. 9–71, Academic Press, London.

Bazin, M.J. and Prosser, J.I. (1992) Modelling microbial ecosystems, *Journal of Applied Bacteriology (Supplement)*, **73**, 89–95.

Booth, I.R., Cairney, J., Sutherland, L. and Higgins, C.F. (1988) Enteric bacteria and osmotic stress: an integrated homeostatic system, *Journal of Applied Bacteriology (Supplement)*, **65**, 35S–49S.

Carpentier, B. and Cerf, O. (1993) Biofilms and their consequences, with particular reference to hygiene in the food industry, *Journal of Applied Bacteriology*, **75**, 499–511.

Cavalier-Smith, T. (1987) Eukaryotes with no mitochondria, *Nature*, **326**, 332–3.

Chesbro, W., Evans, T. and Eifert, R. (1979) Very slow growth of *Escherichia coli*, *Journal of Bacteriology*, **139**, 625–38.

De Graeve, K.G., Grivet, J.P., Durand, M. *et al.* (1994) Competition between reductive aceto-genesis and methanogenesis in the pig large-intestinal flora, *Journal of Applied Bacteriology*, **76**, 55–61.

Evans, H.C. (1988) Coevolution of entomogenous fungi and their insect hosts, *Coevolution of Fungi with Plants and Animals* (eds. K.A. Pirozynski and D.L. Hawksworth), pp. 149–71, Academic Press, London.

Fernandez-Linares, L., Faure, R., Bertrand, J.-C. and Gauthier, M. (1996) Ectoine as the predominant osmolyte in the marine bacterium *Marinobacter hydrocarbonoclasticus* grown on eicosane at high salinities, *Letters in Applied Microbiology*, **22**, 169–172.

Gerber, S. and Hader, D.-P. (1994) Effects of enhanced UV-B irradiation on the red coloured freshwater flagellate *Euglena sanguinea*, *FEMS Microbiology Ecology*, **13**, 177–84.

Goumelon, M., Cillard, J. and Pommepuy, M. (1994) Visible light damage to *Escherichia coli* in seawater: oxidative stress hypothesis, *Journal of Applied Bacteriology*, **77**, 105–12.

Gounot, A.-M. (1991) Bacterial life at low temperature: physiological aspects and biotech-nological implications, *Journal of Applied Bacteriology*, **71**, 386–97.

Herbert, R.A. (1992) The application of micro-electrodes in microbial ecology, *Journal of Applied Bacteriology (Supplement)*, **73**, 164–72.

Hicks, S.J. and Rowbury, R.J. (1987) Bacteriophage resistance of attached organisms as a factor in the survival of plasmid-bearing strains of *Escherichia coli*, *Letters in Applied Microbiology*, **4**, 129–32.

Kaprelyants, A.S., Gottschal, J.C. and Kell, D.B. (1993) Dormancy in non-sporulating bacteria, *FEMS Microbiology Reviews*, **104**, 271–86.

Keith, S.M., Russ, M.A., Macfarlane, G.T. and Herbert, R.A. (1987) The ecology and physiology of anaerobic bacteria isolated from Tay estuary sediments, *Proceedings of the Royal Society of Edinburgh*, **92B**, 323–33.

Kersters, I., Huys, G., van Duffel, H. *et al.* (1996) Survival potential of *Aeromonas hydrophila* in fresh water and nutrient poor waters in com-parison with other bacteria, *Journal of Applied Bacteriology*, **80**, 266–76.

Lee, J.V. and West, A.A. (1991) Survival and growth of *Legionella* species in the environment, *Journal of Applied Bacteriology (Supplement)*, **70**, 121–30.

Lewis, D.L. and Gattie, D.K. (1990) Effects of cellular aggregation on the ecology of micro-organisms, *American Society of Microbiology News*, **56**, 263–8.

Malloch, D.W. (1988) The evolution of mycorrhizae, *Canadian Journal of Plant Pathology*, **9**, 398–402.

Moran, M.A., Rutherford, L.T. and Hodson, R.E. (1995) Evidence for indigenous Streptomyces populations in a marine environment determined with a 16S rRNA probe, *Applied and Environmental Microbiology*, **61**, 3695–3700.

Mouget, J.-L., Pakhama, A., Lavoie, M.C. and de la Noue, J. (1995) Algal growth enhancement by bacteria: Is consumption of photosynthetic oxygen involved?, *FEMS Microbiology Ecology*, **18**, 35–44.

Moyer, C.L., Dobbs, F.C. and Kail, D.M. (1994)

Estimation of diversity and community structure through restriction fragment length polymaplian distribution analysis of bacterial 16S rRNA genes for a microbial mat at an active hydrothermal vent system, Loihi Seamount, Hawaii, *Applied and Environmental Microbiology*, **60**, 871–9.

Muller, R.H. and Babel, W. (1996) Measurement of growth at very low rates, an approach to study the energy requirements for the survival of *Alcaligenes eutrophus* JMP 134, *Applied and Environmental Microbiology*, **62**, 147–51.

Nielsen, T. and Eklund, N.G.A. (1995) Influence of solar ultraviolet radiation on photosynthesis and motility of marine phytoplankton, *FEMS Microbiology Ecology*, **18**, 281–8.

Nojoumi, S.A., Smith, D.G. and Rowbury, R.J. (1995) Tolerance to acid in pH 5.0 grown organisms of potentially pathogenic Gram-negative bacteria, *Letters in Applied Microbiology*, **21**, 359–63.

Oliver, J.D., Hite, F., McDougald, P. *et al.* (1995) Entry into, and resuscitation from the viable but non culturable state by *Vibrio vulnificus* in an estuarine environment, *Applied and Environmental Microbiology*, **61**, 2624–2630.

Ozkanka, R. and Flint, K.P. (1996) Alkaline phosphatase activity of *Escherichia coli* starved in sterile lake water, *Journal of Applied Bacteriology*, **80**, 252–8.

Pirozynski, K.A. and Hawksworth, D.L. (1988) Coevolution of fungi with plants and animals: Introduction and overview, *Coevolution of Fungi with Plants and Animals* (eds. Pirozynski, K.A. and Hawksworth, D.L.), pp. 1–29, Academic Press, London.

Rajendram, N., Matsuda, O., Urushigawa, Y. and Simidu, U. (1994) Characterization of microbial community structure in the surface sediment of Osaka Bay, Japan by phospholipid fatty acid analysis, *Applied and Environmental Microbiology*, **60**, 248–57.

Rehnstam, A.-S., Backman, S., Smith, D.C. *et al.* (1993) Blooms of sequence-specific culturable bacteria in the sea, *FEMS Microbiology Ecology*, **102**, 161–6.

Roszak, D.B. and Colwell, R.R. (1987) Survival strategies of bacteria in the natural environment, *Microbiological Reviews*, **51**, 365–79.

Russell, J.B. (1992) Another explanation for the toxicity of fermentation acids at low pH: anion accumulation versus uncoupling, *Journal of Applied Bacteriology*, **73**, 363–70.

Russell, N.J. (1990) Cold adaptation of micro-organisms, *Philosophical Transactions of the Royal Society of London B*, **326**, 595–611.

Russell, N.J. and Fukunaga, N. (1990) A comparison of thermal adaptation of membrane lipids in psychrophilic and thermophilic bacteria, *FEMS Microbiology Reviews*, **75**, 171–82.

Sutherland, L., Cairney, J., Elmore, M.J. *et al.* (1986). Osmotic regulation of transcription: Induction of the proU betaine transport gene is determined by the accumulation of intracellular potassium, *Journal of Bacteriology*, **168**, 805–814.

Woolfaardt, G.M., Lawrence, J.R., Robarts, R.D. *et al.* (1994) Multicellular organization in a degradative biofilm community, *Applied and Environmental Microbiology*, **60**, 434–6.

2 AQUATIC ENVIRONMENTS

In terms of physical size, the aqueous environment is the largest. The oceans represent the greatest area, but inland lakes and rivers are by no means insignificant. It has, in fact, been argued that the distinction between the oceans and freshwater environments is artificial and that all form part of the same continuum. In practice both oceans and freshwater systems contain a number of related, but distinct environments; conditions in the ocean depths, for example, are very different to those in the surface layers, which again differ from those in estuaries. Equally conditions in a eutrophic inland lake are likely to be very different to those in a low-nutrient mountain stream. With the exception of specific circumstances that permit rapid microbial growth, however, aquatic environments tend to be severe and successful micro-organisms deploy a variety of survival strategies. Strategies are also required to deal with competition from and co-operation with other micro-organisms. Despite these factors many aquatic environments support a wide range of micro-organisms, from phage through eubacteria and cyanobacteria to microfungi, protozoa and eukaryotic algae.

MICROFLORA OF THE DIFFERENT AQUATIC ENVIRONMENTS

Introduction

Aquatic environments range from the pristine stream emerging from the mountain top to the chimneys of hydrothermal vents in the deepest parts of the ocean. At the same time, ecosystem habitats may range in size from microniches on small pieces of particulate matter to stratified lakes and microbial mat communities. It has, however, been argued that, with the exception of specialized micro-organisms, the ecology of marine and freshwater micro-organisms is similar. The extent to which this is true is debatable, but some microbiologists consider that micro-organisms of marine and freshwater origin coexist in estuarine and inshore environments where fresh water and sea water mix. Further evidence is required for a full understanding of the situation. It is established, however, that there are many common features of life in photic aquatic communities where algae and cyanobacteria are of much greater importance in primary production than in most terrestrial environments.

The importance of predation in controlling the bacterial population has long been recognized in both aquatic environments, where hetero-trophic nano-flagellates are important consumers of bacteria and play a major role in the regeneration of nutrients. There is considerable diversity among the hetero-trophic nanoflagellates, and their relationship with the prey population can vary considerably (*Feature 3*).

A role for predatory bacteria in limiting the size of the overall bacterial population has been discussed, but is generally thought to be very limited. The parasitic bacterium *Bdellovibrio bacteriovorus* (**104–106**) is present in seawater and freshwater, but its overall importance is probably limited. *Bdellovibrio bacteriovorus* may, however, be of much greater importance in restricted niches where the prey density is high. The gliding aquatic myxobacter, *Lysobacter* (**107, 108**) is algicidal and may limit the extent of blooms.

The importance of bacteriophage activity (**15–17**) in aquatic environments of all types has only recently been appreciated. The properties of the phage population of aquatic environments are poorly understood but it is considered that in the oceans and probably other waters, viruses are the most numerous micro-organism. Bacteriophages are of considerable ecological importance, not only for limiting the numbers of other micro-organisms but also because nutrients released after lysis are available for direct utilization only by other micro-organisms. In this way nutrient movement upwards through the food chain is short-circuited. It has been argued that bacteriophage activity, while potentially lethal for individual micro-organisms, actually benefits the microbial community as a whole by maintaining nutrient availability. Equally, it has been demonstrated, using fluorescently labelled probes, that viruses can, although not invariably, control the structure of marine microbial communities at levels below one infective unit ml^{-1} (Hennes *et al.*, 1995).

It is also recognized that micro-organisms in aquatic environments are subject to starvation and to long periods at low temperatures. There is, however, a matter of magnitude, e.g. bacteria carried from ocean surface layers to the depths by water upheavals may have to survive cold, starvation and high pressure for many years before returning to more favourable conditions. In contrast, deprivation in freshwater environments may be regarded as minimal and short-lasting.

Feature 3. Feeding strategies

Heterotrophic nanoflagellates are a complex group of organisms that are able to coexist by employing different survival strategies and occupying different ecological niches. This may be demonstrated by the differing bacterial consumption patterns of starved cells of a filter feeder and an interception feeder. The filter feeder *Pteridomonas danica* continuously consumed four to five bacteria per hour over an eight hour period, irrespective of bacterial density above 5×10^5 ml^{-1}. Catching continued at lower bacterial densities. In contrast, the interception feeder *Paraphymonas imperforata* (which has only half the biomass of *Pteridomonas danica*) ingested ten bacteria per hour at high bacterial densities but reduced feeding rate with reducing densities. Feeding ceased at a bacterial density of 1×10^6 ml^{-1}. Data from Zubkov and Sleigh, 1995.

15–17 Morphology of typical viruses isolated from an aquatic (lake) environment.

Blooms

Blooms are a feature of both seawater and freshwater environments. Algal and cyanobacterial blooms are well recognized, but less attention has been paid to bacterial blooms.

Algal and cyanobacterial blooms

Algal blooms occur as a consequence of eutrophication, an increase in nutrient load, of bodies of both fresh and salt water. Algal blooms appear to be increasing in frequency, at least partially as a consequence of man's activities, especially increased levels of phosphorus and nitrogen. Algal blooms are natural occurrences and the spring phytoplankton bloom in temperate and arctic seas is a well-known phenomenon. Blooms are predictable consequences of hydrological changes which increase the nutrient content of upper layers of water. These include run-off from land, water column turnover and up-welling, bringing deep, nutrient-rich water to the surface. The spring bloom results from the onset of stratification and a high nutrient level due to vertical mixing, coinciding with increased day length and quantity of solar radiation.

Under favourable conditions, the growth of phytoplankton is exponential. Algal biomass and bloom formation may be controlled by herbivore grazing but response is slow, permitting numbers to rise rapidly to $c.10^8$ ml^{-1}. Numbers of algal phages increase rapidly with increasing cell numbers and metabolic activity. The role in preventing bloom formation is not known but on at least one reported occasion a bloom has been stopped by phage. Species succession occurs as conditions change but competition means that, at their peak, blooms are usually monospecific, *Phaeocystis* for example being most common in spring blooms in coastal North Sea zones. Blooms are often recognized visually by discoloration of the sea which may be green, brown, or red depending on the algae present. White blooms are due to the calcareous skeletal elements of coccolithophores, such as *Emiliania huxleyi*, while 'red tides' are caused by dinoflagellates.

The high levels of organic material produced by blooms can have beneficial effects further up the food chain though, in many cases, production of organic matter is so great, that remineralization by bacteria consumes very large quantities of dissolved oxygen. This, together with oxygen consumption by the night-time respiration of algae, leads to anoxia and mass mortality amongst fish and invertebrates. Fish may also be killed by toxins and by H_2S production. The effect of blooms is often particularly marked in closed systems such as lakes; long-term effects on ecosystems often result. Cyanobacterial blooms reduce light which may ultimately lead to the disappearance of larger aquatic plants and visually hunting piscivorous fish. Filter-feeding planktonivorous fish become dominant leading to the virtual disappearance of crustacean zooplankton, reducing the grazing pressure on algae.

From the viewpoint of man, blooms are undesirable for a number of reasons. They are foul smelling and their presence results in a loss of recreational value in both coastal and inland waters. A number of genera produce toxins which may either directly affect man and animals entering the water, or indirectly through consumption of shellfish that have acquired algal toxins. Not all algae involved in bloom formation are toxigenic but toxins may be involved in antipredator strategies and toxigenic species are favoured under conditions supporting bloom formation. In recent years some toxigenic algae have spread geographically. *Gymnodinium catenatum*, for example, has spread from southern Californian waters to southern Europe, Japan and Australia. Also, the previously non-toxigenic genus *Nitzschia pungens* is now recognized as producing a toxin – domoic acid – and is associated with amnesic shellfish poisoning.

Bacterial blooms

Bacterial blooms are less visible than algal or cyanobacterial blooms and have a less detrimental effect on the environment. Bacterial blooms occur in near-shore and freshwater environments under similar conditions to algal and cyanobacterial blooms.

The role of aquatic micro-organisms in the formation of carbonate minerals

A number of micro-organisms, of which cyanobacteria are of considerable importance, are involved in the formation of carbonate minerals. Initial work suggested that this phenomenon was confined to marine systems but cyanobacteria are now known to be essential mediators of calcite formation in freshwater environments (Schulzte-Lam and Beveridge, 1994).

In water containing abundant Ca^{2+} ions and dissolved CO_2 – primarily as HCO_3^- in alkaline waters – cyanobacteria provide nucleation sites for mineral formation to commence. Utilization of HCO_3^- leads to the release of OH^- ions and a rise in pH value in the microenvironment surrounding the cell. The formation of carbonate minerals is highly pH dependent and the degree of alkalization determines the mineral phase developing. Carbonate phases are deposited at pH values in excess of 8.0 while other mineral phases are formed at lower pH values.

Studies of a meromictic lake of high Ca^{2+} and SO_4^{2-} content showed gypsum ($CaSO_4 . 2H_2O$) to be initially formed on cells of *Synechococcus*. As cellular activity increased, however, the dominant phase became calcite. Calcitic microbial reefs were present on the lake shore and marl sediments deposited.

It is now known that biogenesis of other carbonate minerals, which were previously considered to be of non-biological origin, is possible (Schulzte-Lam and Beveridge, 1994). These include strontianite ($SrCO_3$) and magnesite ($MgCO_3$).

THE MARINE ENVIRONMENT

Introduction

Conditions within the marine environment as a whole vary widely and it is necessary to discuss the various major parts separately. It is customary to categorize the various types of micro-organism biologically and by size. Thus the bacterio- or pico-plankton are in the size range 0.2–2.0 µm, much of this population being ultramicrobacteria. The phytoplankton (plants) is subdivided into nano-plankton (2.0–20 µm) and microplankton (20–200 µm). Microscopic animals are zooplankton and bacteriophages femtoplankton.

Many of the bacteria from marine environments are characterized by an obligate requirement for Na^+ and other ions in seawater. Some of these are closely related to freshwater bacteria, which have no such requirement or a requirement at a significantly lower level. Chitinolytic activity is extremely common amongst marine bacteria as a consequence of the vast supplies of chitin available from the bodies of marine arthropod exoskeletons. Agarolytic activity is also widespread, especially amongst marine cytophages.

The dissolved organic matter (DOM) present in seawater is one of the largest pools of organic matter in the biosphere. Dissolved organic matter is a major substrate for heterotrophic bacteria and an intermediate in the carbon flux from phytoplankton and zooplankton to bacteria. Only a discrete fraction is available as substrate for heterotrophs, the remainder being highly recalcitrant with a turnover time which may be as long as several thousand years. Dissolved humic matter (DHM) accounts for most of this recalcitrant material. This material has a similar $^{13}C:^{12}C$ ratio to phytoplankton and is distinct from humic material derived from terrestrial plants. It therefore appears that a substantial quantity of recalcitrant DOM, including DHM, is a product of primary production (Tranvik, 1993). Dissolved humic matter probably comprises the recalcitrant fraction of the DOM together with complex, recalcitrant matter produced from the labile fraction of DOM, possibly as a consequence of sequential assimilation and excretion by several different micro-organisms.

The ocean

As a consequence of its great depths, the ocean consists of several broad groups of habitat of differing physicochemical characteristics (18). These may be grouped into the sea surface microlayer, inhabitants of which are often categorized as the neuston. Surface waters above the thermocline, which are illuminated and heated by solar radiation, comprise the epipelagic habitat, while immediately below the thermocline is the mesopelagic habitat. This merges with the bathypelagic at a depth of c.2,000 m. The average depth of the ocean bed is 3,800 m, below which the habitat is described as abyssopelagic. At an even greater depth are hadal habitats, which extend below 11,000 m.

The sea surface microlayer

The sea surface microlayer is a thin film of liquid at the interface of the sea and the atmosphere, which is of considerable importance in exchanges between sea and atmosphere. The film is oily, consisting primarily of lipids derived – in the absence of pollution – from external sources, from marine organisms and natural hydrocarbon seeps.

18 Classification of habitats in relation to the topography of the ocean. All habitats in the water column are pelagic while sea bed habitats are benthic. Both pelagic and benthic habitats in the continental shelf region are included in the neritic habitat.

18
---------- High water ----------------------
Littoral
---------- Low water ----------------------
Epipelagic (photic)
---------- Thermocline ----------------------
Mesopelagic
---------- 2,000 m ----------------------
Bathypelagic
---------- 3,800 m ----------------------
Abyssopelagic
---------- Oceanic trenches ----------------------
Hadal
---------- 11,000 m ----------------------

Feature 4. The Great Magic Trick

Cyanobacteria, such as *Trichodesmium, Nostoc, Nodularia* and others, are of very significant ecological importance in oligotrophic waters as a consequence of their ability to fix both carbon and nitrogen. The 'Great Magic Trick' for these micro-organisms is to manage both oxygenic photosynthesis and nitrogen fixation, the latter being an extremely oxygen-sensitive process. The basic problem is to protect the highly oxygen-sensitive enzyme nitrogenase from both endogenously evolved oxygen and that in the environment (Fay, 1992).

Different cyanobacteria have adopted different management strategies to the problem of both oxygenic photosynthesis and nitrogen fixation. *Plectonema*, for example, has adopted a very simple strategy of fixing nitrogen only under conditions of low oxygen (micro-oxic) tension. A further simple strategy is that of cyanobacteria such as *Oscillatoria limnosa* which fix nitrogen only during the dark, non-photosynthetic hours of the night. The most sophisticated approach appears to lie with those cyanobacteria that have evolved heterocysts – specialized anaerobic cells lacking oxygen-evolving mechanisms and which have additional cell wall layers to minimize oxygen penetration. Cyanobacteria of this type include *Anabaena, Nostoc* (**25**), *Nodularia* (**26, 27**) and *Aphanizomenon* (**28**).

Trichodesmium, however, is the greatest magician of all! This organism has the ability to fix nitrogen during the day while lacking the specialized oxygen-protective heterocysts. It has been extensively studied at the University of Stockholm using immunogold/transmission electron microscopy and immunofluorescence/light microscopy. Examination of three species of *Trichodesmium* (**29–31**) showed nitrogenase to be restricted to a limited (c.14%) subset of cells randomly distributed throughout the colony. The frequency of nitrogenase-containing cells varied diurnally, being highest during the day, the period of active nitrogen fixation and falling overnight to approach zero just before dawn (**32–36,** see page 38). It appears that colonies of *Trichodesmium* contain specialist nitrogen-fixing cells or whole filaments. Segregation of nitrogenase to specialist cell types may be seen as analogous to heterocyst production and the pattern of nitrogen fixation (only during daytime and with a peak at noon) is the same as heterocystous cyanobacteria. Although a number of theories have been advanced it is not known how the nitrogenase is protected from oxygen.

Data based on Fredriksson and Bergman, 1995.

With the exception of equatorial and subarctic areas of the Pacific ocean, open epipelagic waters are nitrogen deficient. Despite this, N_2 fixation by either cyanobacteria or N_2-fixing organotrophs is unusual. In these circumstances the non-heterocystous cyanobacterium *Trichodesmium* (**24**), which is capable of concurrent N_2 and CO_2 fixation, is of major importance in primary nitrogen and carbon cycling. *Trichodesmium* is a colony-forming cyanobacterium occurring over very large areas of tropical and subtropical oceans (*Feature 4*). It is stimulated by iron, the loss of which through sedimentation is an acute and chronic problem. The

24 Diagrammatic representation of cells of *Trichodesmium*, a filamentous cyanobacterium. Filaments often develop as parallel bundles of 'trichomes'.

25 Heterocystous cyanobacteria. *Nostoc* is involved in symbiotic relationships, the strain illustrated having been isolated from *Gunnera monoika. Nodularia* has a regular pattern of heterocysts throughout the filament.

26 Heterocystous cyanobacteria. A strain isolated from the Baltic Sea.

27 Heterocystous cyanobacteria. Autofluorescence; there is no fluorescence from the heterocysts. In contrast to *Nodularia*, *Aphanizomenon* produces few heterocysts (one or less per filament).

28 Heterocystous cyanobacteria. A Baltic Sea isolate is illustrated, showing autofluorescence even from heterocysts. **25–28** reproduced by courtesy of Ulla Rasmussen, Department of Botany, University of Stockholm, Sweden.

29 One of three species of *Trichodesmium* representing the major fusiform species in the Caribbean. Colonies, which have been cross-sectioned and stained with toluidine-blue, were classified primarily according to cell size and gas vacuole (arrowed) arrangement. *Trichodesmium thiebautii* has the largest cell diameter of the three species and a number of gas vacuoles scattered over the cells.

30 The second of the three species, *Trichodesmium tenui*, produces a large number of filaments in each colony and is further characterised by one, or a few, gas vacuoles per cell.

31 The third of the three species, *Trichodesmium erythraeum*, is markedly less common in the Caribbean than *Tr. thiebautii* and *Tr. tenui*. The organism is characterised by the distinct cylindrical gas vacuole 'tube' in the cell periphery. **29–31** reproduced with permission from Fredriksson and Bergman, 1995, © 1995 The Society for General Microbiology.

sediments and even in the chimneys themselves. Growth is sufficient to support invertebrate food chains including filter feeding by clams and mussels, grazing by fish and symbioses (see below). Such chains ultimately depend on O_2 from photosynthesis for aerobic chemosynthesis. Chemosynthesis can, however, proceed independently of photosynthesis while combined nitrogen is supplied by free-living nitrogen fixers.

Hyperthermophilic environments exist in black smokers at temperatures of 80–110°C (176–230°F). Black smoker bacteria are usually strict anaerobes due to the low solubility of O_2 in water at high temperatures. Thermophilic, aerobic, non-endo-spore-forming, heterotrophic bacteria have, however, now been isolated from hyperthermophilic environments in deep-sea vents (Marteinsson *et al.*, 1995). Bacteria from such environments exhibit considerable adaptation and cannot usually grow at temperatures below 60°C (140°F) although survival in cold, oxic water is possible for at least a year. Most strains are sulphidogenic, oxidizing H_2 with sulphur, sulphate and CO_2 and fixing CO_2. Heterotrophic growth may also be possible. Major metabolic reactions are summarized in **41**.

Animals and plants are unable to grow at temperatures above *c.*40°C (104°F) and are absent from the actual sites of hydrothermal vents. A very steep thermal gradient exists at vent sites, however, and animals are able to live in close proximity. Symbioses have developed between these deep-sea animals and chemolithotrophic bacteria living in the tissues. The first symbioses studied involved the utilization by bacteria of reduced energy sources, such as H_2S, present in hydrothermal fluids for autotrophic CO_2 formation. The hosts – tube worms, clams and mussels – benefit through an internal source of nutrients while the bacterial symbionts receive inorganic substrates for chemosynthesis. At a later date, a symbiotic relationship was discovered between deep-sea invertebrates dwelling in the region of cold, methane seeps and methophilic bacteria. This symbiosis was initially thought to be restricted to cold seeps. Investigations at the 'Snakepit', an active hydrothermal site in the mid-Atlantic ridge, however, showed methophilic symbioses between bacteria and macrofauna rather than sulphur-oxidizing, autotrophic symbioses (Cavanaugh *et al.*, 1992). Methophilic symbioses involve a simple, short and efficient flow of carbon from methane to the bacterium and from the bacterium to the animal. This may have biotechnological potential (Taylor, 1992).

Neritic and littoral habitats

Neritic (near-shore) and littoral (intertidal) habitats differ in a number of ways from oceanic habitats. Littoral habitats, in particular, are stressful environments for both marine and non-marine microorganisms. Despite this, turnover of organic compounds is much more intense and about 90% of bacterial degradation occurs in the benthic zone of the continental shelf region, which occupies only 10% of the total marine environment. Nutrients are available at a higher level than in ocean habitats and a wider range of organic substrates are present. These are metabolized by heterotrophic bacteria which are also essential for degradation of pesticides and other pollutants.

High light intensities lead to the extensive development of benthic plant communities, including seaweeds, marshgrasses and seagrasses. In tropical and subtropical climates, mangroves and coral reefs are also of importance. Primary production by plants is at a high level, enhanced by rapid regeneration of inorganic nutrients and input of nutrients from terrestrial sources. Bacteria are heavily involved in biogeochemical cycles: carbon, nitrogen, sulphur, phosphorus and trace metals. Combined nitrogen is recycled by the degradative activities of heterotrophic bacteria. Cyanobacteria develop very extensive mat communities which are of major importance both in nitrogen fixation and photosynthesis. A highly diverse range of prokaryotic micro-organisms in mats are capable of nitrogen fixation and the distribution of nitrogenase genes indicates that the ability to fix nitrogen cannot be predicted on the basis of taxonomy. A

41

Chemolithotrophic

$$H_2 + S° = H_2S$$

$$4H_2 + SO_4^{2-} = S^{2-} + 4H_2O$$

$$CO_2 + SO_4^{2-} = CH_4 + 2H_2O$$

Heterotrophic

$$CH_2O + 2S° + H_2O = CO_2 + 2H_2S$$

$$2CH_2O + SO_4^{2-} = S^{2-} + 2CO_2 + 2H_2O$$

41 Metabolism of hyperthermophiles.

study of nitrogen fixation in mats off the coast of North Carolina showed a seasonal shift, cyanobacteria dominating nitrogen fixation in summer and heterotrophic bacteria dominating in winter (Zehr *et al.*, 1995). Cyanobacteria on the leaves of seagrass also fix significant quantities of N_2 while *Azotobacter* forms a nitrogen-fixing endosymbiosis with the green seaweed *Codium* sp. Sulphate-reducing bacteria are also involved in N_2 fixation in sediments.

Genera of micro-organisms present in the water column of the ocean are also present in neritic and littoral environments. In some cases a wider range of species is present, although significant temperature differences between summer and winter may be reflected in considerable seasonal variation. *Vibrio* is more prevalent than in open oceans and species, including the human pathogens *V. cholerae* and *V. parahaemolyticus* (**42**). Both enter a viable, nonculturable stage during cold conditions, although survival of *V. cholerae* may also be linked with its association with chitin particles. *V. parahaemolyticus* has been described as a 'marine enteric organism' and is often associated with shellfish such as clams. This has public health implications as such products are often eaten raw. In recent years a restricted *V. parahaemolyticus* biovar (O4:K12, urease+) has become dominant in coastal waters of the western US and Mexico and is also the major pathogenic strain. Dominance may reflect a more effective response to environmental pressures (Varnam and Evans, 1991). A further pathogenic species, *V. vulnificus*, which is also associated with shellfish, is amongst vibrios isolated from the stomach of finfish where the pH value is *c.*4.0. Numbers of up to 10^6 g^{-1} have been isolated from bottom-feeding fish consuming molluscs and crustacea. The presence of *V. vulnificus* in finfish is not considered to be of major public health significance, since such fish are cooked before consumption. Finfish may, however, be of importance in the introduction of *V. vulnificus* to new areas. Finfish are mobile, travelling in large schools and could have an immediate effect on water quality in shallow areas with little water movement (DePaola *et al.*, 1994).

Luminescent bacteria are also more widespread in neritic and littoral waters and the luminescent symbiosis more common. Luminescence is a relatively widespread trait in marine bacteria, which may be considerably more common than currently appreciated. The majority of marine luminescent strains are fermentative, Gram-negative rods of the

42 *Vibrio parahaemolyticus*. With the exception of *V. hollisae*, most vibrios grow well on TCBS cholera medium (originally designed for the recovery of *V. cholerae*). Sucrose-fermenting species, including *V. cholerae*, produce yellow colonies while non-sucrose-fermenting species, including *V. parahaemolyticus*, produce green colonies. (Source: seawater; Medium: TCBS cholera medium; Incubation: 37°C (98.6°F), 24 hours.)

43 *Vibrio fischeri*. Like *V. parahaemolyticus*, *V. fischeri* does not ferment sucrose. The strain illustrated produces a yellow-brown pigment that obscures the green colour of the colonies. (Source: inshore seawater; Medium: TCBS cholera medium; Incubation: 37°C (8.6°F), 24 hours.)

genera *Vibrio* and *Photobacterium*, the sole exception being the non-fermentative *Alteromonas hanedai*. The best known luminescent bacteria are *V. fischeri* (**43**), *P. phosphoreum* and *P. leiognathi*, which form symbioses with fish and other marine animals. These bacteria are also free-living in the marine environment and present on the surface of fish as saprophytes. In addition no symbiotic relationships are known for common luminescent bacteria such as *V. harveyi*. In bioluminescent symbioses, the bacterial symbiont is harboured, in pure cultures, in specialized light organs and receives nutrients from the fish. The bacterium also benefits from exclusion of competitors and dissemination from the light organ into new

44 This illustration and the following two show the autoinduction of luminescence and luciferase synthesis in *Vibrio fischeri*. Although cell division recommences c. 1 hour after transfer of luminescent cells to a new culture, luciferase synthesis ceases for c. 4 hours.

45 Cessation of luciferase synthesis leads to a marked decrease in luminescence during the first 4–5 hours after transfer.

47 Structure of the luminescence autoinducer *N*-(3-oxo-hexanoyl) homoserine lactone.

46 A high level of *de novo* synthesis of luciferase after 4–5 hours, when cell numbers are sufficient to permit accumulation of the autoinducer (**47**) to threshold levels. This leads to a dramatic increase in luminescence (**45**).

habitats, such as seawater and the fish gut. Light produced by the bacterium is utilized by the fish in a number of ways, including feeding and, possibly, signalling and antipredation.

Available evidence strongly suggests that life in the light organ for the bacterium involves a low growth rate, 20 to 30 times slower than in laboratory culture, and high light production. In *V. fischeri* luciferase synthesis and the resulting luminescence are controlled by a regulatory mechanism known as autoinduction (**44–46**). Control is mediated by an autoinducer, *N*-(3-oxo-hexanoyl)homoserine lactone (**47**), which acts as a bacterial pheromone. Synthesis of luciferase and other luminescence-associated enzymes is triggered by the concentration of autoinducer reaching a threshold level, usually when cell numbers exceed 10^7 ml^{-1}. This means that the expression of luminescence in *V. fischeri* is frequently coupled with symbiosis since the autoinducer would not be expected to accumulate to threshold levels in seawater where cell densities are typically less than 10^2 ml^{-1}. The mechanism also ensures that the energy hungry process of bioluminescence is avoided, and energy conserved, when cell numbers are low.

48 A halotolerant species of *Bacillus*. Although *Bacillus* is a ubiquitous organism, relatively little is known of isolates from marine sources. This isolate, which resembled the little-studied '*B. filicolonicus*', was capable of growth at NaCl concentrations in excess of 12%. (Source: seawater, English channel; Medium: blood agar; Incubation: 28°C (82.5°F), 24 hours. Note that the original isolation was made on seawater agar but colonial morphology is best seen on blood agar.)

49 Halotolerant *Bacillus* species. The positive Gram reaction has been virtually entirely lost. (Source: as for **45**; Microscopy: Gram stain, × 1,200.)

50 A gliding bacterium from marine mud. Addition of crystal violet to 3.5% NaCl medium to inhibit halotolerant *Bacillus* permitted development of a mixed microflora of Gram-negative, rod-shaped bacteria, including small, swarming colonies. Swarming was unaffected by the relatively high nutrient content of the medium or by the presence of crystal violet, but a minimum of 1% NaCl was required. (Source: intertidal mud flats; Medium: nutrient agar with 0.02% crystal violet and 3.5% NaCl; Incubation: c.18°C (64.4°F), 5 days.)

The environmental benefit of luminescence to free-living or saprophytic cells is less readily apparent. It may be a factor in enhancing dispersal since luminescent micro-organisms, growing saprophytically on dead fish, attain sufficient cell density to induce luminescence. This in turn is likely to attract marine organisms to the food source, resulting in propagation and dispersal (Meighen, 1992). However, non-luminescent bacteria appear to be equally successful in dispersal. A possible direct role for the luminescent system is the function of luciferase as an alternative electron carrier to the cytochrome system. This may enhance survival and growth at low oxygen tensions (Ulitzur *et al.*, 1981).

Some of the genera present in, and adapted to, coastal and estuarine waters are not present in oceans. These include *Bacillus* (**48, 49**), the micro-aerophil *Oceanospirillum*, appendaged bacteria including *Caulobacter*, unicellular gliding bacteria including *Cytophaga*, *Capnocytophaga* and strains of unknown identity (**50**). *Bdellovibrio* has been isolated from coastal seawater, but its significance is unknown. The gliding and highly differentiated bacterium *Leucothrix* is present as an epiphyte on seaweed, especially in intertidal zones. It is present in large numbers in decaying seaweed and important in its degradation. Marine strains of *Cytophaga* are also common and of importance in the degradation of organic matter, comprising up to 33% of the cellulolytic microflora. *Streptomyces* can also be readily isolated from coastal seawater, especially where nutrient concentrations are high.

The presence of this organism was, for many years, attributed to exospores being washed in from terrestrial sources but there is now evidence of an indigenous, marine origin (Moran *et al.* 1995).

Relatively little is known of micro-aerophilic bacteria in seawater, although both neritic and ocean waters would appear to offer favourable conditions for their growth. A study of the microflora of Tokyo Bay showed micro-aerophils to be present at a level of 50–100% of that of aerobes and facultative anaerobes, suggesting a role in mineralization. Nine phena were identified, resembling, but distinct from, non-marine genera (Sugita *et al.*, 1993). Magnetotactic, micro-aerophilic results in S^{2-}, generated at depths of 20–30 mm in the sediment, being made available to phototrophic purple sulphur bacteria resident at a much shallower depth of 1.5–2.0 mm. The cyclic nature of S^{2-} availability leads to the formation of blooms of *Thiocapsa* over a relatively short time scale (**60**, **61**) (van Gemerden *et al.*, 1989).

In sediments of lower organic input, the O$_2$ and H$_2$S zones do not overlap and are separated by a sub-oxic, or Fe–Mn zone. This sub-oxic zone is the site of anaerobic oxidation of organic compounds by Fe(III) and, less importantly, Mn(IV). H$_2$S reacts with Fe(II) to generate FeS which forms pyrites (FeS$_2$) by reaction with polysulphide. Pyrites is re-oxidized by O$_2$ on a seasonal basis. In coastal sediments, re-oxidation occurs in winter when O$_2$ penetrates deepest due to the reduced settlement of detritus. In contrast, re-oxidation in salt marshes occurs in summer when O$_2$ flows into the sediments, promoted by the higher level of photosynthetic activity. Pyrite accumulates in winter.

Methanogenesis is limited in marine environments by competition from sulphate-reducing bacteria. Methanogens, unlike sulphate-reducing bacteria, are able to metabolize trimethylamine derived from the osmolytes trimethylamine oxide and betaine. This pathway permits a certain level of methanogenesis, although in some cases it has been suggested that methanogenesis proceeds at a higher level than can be accounted for in this way. High levels of sulphate reduction occur in some situations including significant sewage pollution. Sulphate-reducing

60 Development of blooms of *Thiocapsa roseopersicina* on a beach.

61 Development of blooms of *Thiocapsa roseopersicina* on a beach. **60** and **61** reproduced by courtesy of Prof. R.A. Herbert, University of Dundee.

bacteria develop in biofilms and a high level of activity leads to rapid corrosion of metal structures. The death of fish and marine animals may also result from incomplete oxidation of high levels of H_2S.

Bacteria in sediments use electron acceptors in a sequence dictated by energy yields. The sequence of oxidation is approximately reflected by the depths within the sediment at which different processes proceed at maximum rates (**62**). This sequence can, however, be disturbed by seepage of groundwater containing for example O_2 and NO_3 beneath the sulphate-reducing zone.

Although general patterns of the microbiology of neritic and littoral zones may be defined, it is important to appreciate that these areas support a wide range of habitats with specialized microfloras. On coral, for example, a horizontally migrating cyanobacterial mat develops. The mat is responsible for the economically important 'black band' disease of coral (**63**) (Carlton and Richardson, 1995).

62 Distribution of electron acceptors for bacterial metabolism in sediments.

62

Water

1.	Organic C + O_2	$\rightarrow$	$CO_2 + H_2O$
2.	Organic C + NO_3^-	$\rightarrow$	$CO_2 + N_2$
3.	Organic C + NO_3^-	$\rightarrow$	$CO_2 + NH_4^-$
4.	Organic C + MN(IV)	$\rightarrow$	CO_2 + MN(II)
5.	Organic C + Fe(III)	$\rightarrow$	CO_2 + Fe(II)
6.	Organic C + SO_4^-	$\rightarrow$	$CO_2 + S$
7.	Organic C	$\rightarrow$	$CO_2 + CH_4$

63 Coral is of geological importance in favouring the development of atolls and protecting coastlines. It is a habitat for a range of life, including reef dwelling fish. Some types of coral are used to make decorative items and can be of economic significance to local communities. 'Black band' disease is aptly named, the band being seen clearly between the lower living coral and the upper dead material (the nail provides a perception of scale). The cause appears related to sulphide production by a cyanobacterial mat. This resembles other cyanobacterial mats in structure but is unusual in migrating horizontally rather than vertically. Reproduced with permission from Carlton and Richardson, 1995, © Federation of European Microbiological Societies.

Consequences of salinity changes

A feature of neritic and littoral habitats is that the salinity is subject to marked and, in some cases, sudden variation. For this reason environments of this type offer a considerable challenge both to marine and freshwater micro-organisms although the two 'types' are able to coexist. The particular circumstances affecting micro-organisms are the consequence of the proximity of the terrestrial environment and fresh water entering the sea through rivers, drains, and flooding with consequent localized dilution. Estuarine environments are particularly subject to changes in NaCl concentration through changes in river flows, tidal patterns, and the like. At the same time, evaporation of water in hot weather means that NaCl concentrations in localized parts of intertidal habitats can rise significantly under some tidal conditions.

Protection against osmotic stress is obtained by the accumulation of a restricted range of low molecular weight organic compounds (compatible substrates). *Thiocapsa* from marine mats, for example, accumulates K^+ and sucrose. Glycine betaine is important in halotolerant and halophilic bacteria but is not accumulated by other bacteria. There is also evidence that slime offers protection, probably through maintaining a microenvironment around the cell and preventing sudden osmotic change (64, 65).

Most of the bacteria and cyanobacteria present in environments of fluctuating salinity are able to grow over a wide NaCl range (66). Some may originally be of terrestrial origin (see above) while many marine bacteria have only a very low requirement for Na^+ or other ions. There may, however, be physiological or morphological changes (67, 68, 69) as a consequence of growth at different

64 Effect of osmotic shock on a marine *Vibrio* species. The organism, isolated from seashore mud, produces copious slime when grown in liquid culture, adhering strongly to the walls of the culture vessel and increasing the viscosity of the medium. This is reflected in the appearance of colonies on seawater agar.

65 Cells (1) transferred from a medium containing 4% NaCl to water retained both integrity and viability. In contrast, cells (2) from a mutant, non-slimy strain of the same *Vibrio* lysed immediately on transfer to water.

66 An unidentified, halophilic, Gram-negative, rod-shaped bacterium. The bacterium, which differed from known genera, was able to grow at NaCl concentrations ranging from 0.5 to more than 25% added NaCl. The optimum is 7.5–10%. (Source: intertidal muddy pool; Medium: tryptone–soya agar + 7.5% NaCl; Incubation: c.18°C (64.4°F), 7 days.)

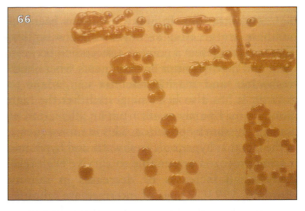

67 Effect of NaCl concentration on cellular morphology of an unidentified, halophilic, Gram-negative, rod-shaped bacterium. At NaCl concentrations of 7.5–25% the cells are fairly regular although the arrangements of some groups superficially resemble those of 'coryneform' bacteria. (Source: as for **66**; Microscopy: phase contrast, × 1,100.)

68 Effect of NaCl concentration on cellular morphology of an unidentified, halophilic, Gram-negative, rod-shaped bacterium. At lower NaCl concentrations than in **67** (5.0%), a few isolated cells of extreme length are present. (Source: as for **66**; Microscopy: phase contrast, × 1,100.)

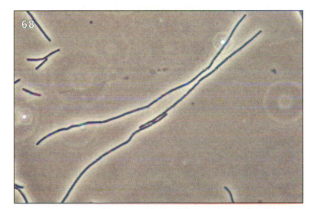

69 Effect of NaCl concentration on cellular morphology of an unidentified, halophilic, Gram-negative, rod-shaped bacterium. At 2.5% NaCl, the majority of the cells are of extreme length. Despite the elongation there are no other indications of ionic stress affecting the cells. (Source: as for **66**; Microscopy: phase contrast, × 1,100.)

able to multiply in moving water develop preferentially as biofilms although there is also colonization of particulate organic matter and aquatic plants. Many of the bacteria present in the water current have been shed from biofilms and are able to rapidly colonize new surfaces (72, 73). A wide variety of bacteria are present in freshwater including species of *Pseudomonas* (74–79), *Flavo-*

72 Colonization of a twig, freshly fallen into a small stream, commenced at points where flow was less strong.

73 Colonizing bacteria were those most common in the general aquatic environment.

74 *Pseudomonas fluorescens*. Many strains of *Ps. fluorescens* isolated from water, soil and foods in Europe are non-pigmented. Most US isolates, however, produce fluorescent pigments. (Source: mountain stream; Medium: quarter-strength nutrient agar; Incubation: 22°C (71.6°F), 5 days.)

75 *Pseudomonas fluorescens*. A typical Gram-negative, rod-shaped bacterium; the cells of this strain are larger than most. (Source: as for **74**; Microscopy: Gram stain, × 1,200.)

76 *Pseudomonas aeruginosa* is common in natural waters although usually present in small numbers. The ability of this organism to grow at relatively high temperatures, and its resistance to antimicrobial agents, enables it to attain high numbers (up to 10^8 cfu ml^{-1}) in jacuzzis and whirlpool baths. Pigment production is slow and not visible in the young culture under the lighting conditions used. Pigment production increases as senescence approaches and colonial appearance becomes highly characteristic. Appearance is usually related to strain. (Source: lowland river; Medium: nutrient agar: Incubation: 35°C (95°F), 24 hours.)

bacterium (80, 81), *Alcaligenes*, *Cytophaga* and 'coryneforms'. These bacteria are consistently present in significant numbers in running water of all types. Many are pigmented, including members of genera normally considered non-pigmented. As in the marine environment, pigmentation is considered to impart protection against UV irradiation. The presence of dissolved organic compounds in freshwater also minimizes the effects of UV irradiation. Members of the *Enterobacteriaceae* can also be isolated from running waters in significant numbers. These bacteria have been associated with

77 *Pseudomonas aeruginosa*. (Source: lowland river; Medium: nutrient agar: Incubation: 35°C (95°F), 24 hours.)

78 *Pseudomonas aeruginosa*. (Source: lowland river; Medium: nutrient agar: Incubation: 35°C (95°F), 4 days.)

79 *Pseudomonas aeruginosa*. (Source: lowland river; Medium: nutrient agar: Incubation: 35°C (95°F), 4 days.)

80 A putative *Flavobacterium*. Although many aquatic isolates are yellow pigmented, this isolate was unusual in that the slime was pigmented but not the colony itself (cf. **81**). Visible slime was not produced until the colony was fully developed. (Source: water, emerging through slime layer at outlet of clay drainpipe; Medium: quarter-strength nutrient agar; Incubation: 22°C (71.6°F), 6 days.)

81 A putative *Flavobacterium*. The isolate was phenotypically similar to, but distinct from, **80**. The colony itself is pigmented but slime and spreading growth at the periphery is, initially, colourless. (Source: ditch receiving water from land drain; Medium: quarter-strength nutrient agar; Incubation: 22°C (71.6°F), 6 days.)

faecal pollution but this is a misconception in the case of environmental strains including *Entero-bacter*, *Hafnia* and *Serratia* (82) which should be considered as a part of the normal microflora of freshwater. *Proteus* (83, 84) is also common and may be present in high numbers in water of high organic content. Some members of the *Entero-bacteriaceae*, such as *Yersinia intermedia* (85), may originally be of animal origin but are able to adapt to aquatic life. Some members of the normal microflora are present only in small numbers and may require special techniques for isolation. These include the Actinomycetes, *Actinoplanes* (86) and *Micromonospora* (87–89) and the gas vacuolated bacterium *Ancyclobacter aquaticus* (*Microcyclus aquaticus*) (90).

82 A prodigiosin-producing strain of *Serratia marcescens*. Pigmented strains of *S. marcescens* may synthesize either the non-diffusible red pigment prodigiosin or the diffusible pink pigment pyrimine. Non-pigmented strains are also common and form the majority of isolates from clinical material. (Source: chalk stream, lightly polluted by agricultural activities; Medium: nutrient agar; Incubation: 30°C (86°F), 2 days.)

83 A species of *Proteus*. The strain swarms on solid media but the periodicity normally associated with swarming in this organism is absent. (Source: silted side channel, lowland river; Medium: nutrient agar; Incubation: 30°C (86°F) , 2 days.)

84 A species of *Proteus*. (Source: as for **83**; Microscopy: Gram stain, × 1,200.)

85 *Yersinia intermedia*. 'Environmental' strains of *Yersinia* are not uncommon in fresh water, although enrichment is usually required for their isolation. The pathogen *Y. enterocolitica* has also been isolated from water although its presence is thought to be due to contamination from animal sources, especially pigs and rats. (Source: lowland, moderately polluted river; Medium: cephaloridin–irgasan–novobiocin agar; Incubation: 28°C (82.4°F), 24 hours.)

86 The sporangium of *Actinoplanes* arises by growth and coiling of the sporangial mycelium within the sporangium wall. The process is thus analogous to that of the streptomycetes, the sporangial wall serving the same function as the fibrous sheath of the streptomycete aerial mycelium. Aerial mycelia appear to be advantageous only in terrestrial habitats when dispersal of spores is by air. The mechanism of spore-bearing in *Actinoplanes* and *Micromonospora* is presumably better adapted to dispersal in water.

87 A species of *Micromonospora*. The organism (arrowed) is present in large but apparently variable numbers in river systems. Recovery can be enhanced by pretreatment of the sample at 80°C (176°F). Colonies can be mistaken for moulds in young cultures. (Source: river water, stored for four weeks at ambient temperature; Medium: nutrient agar: Incubation: 25°C (77°F), 5 days.)

89 *Actinoplanes* and other actinoplanetes form from one to more than 1,000 spores. These are usually motile, within a sporangium arising directly from the substrate mycelium.

88 Although rRNA sequencing indicates that *Micromonospora* is closely related to other actinoplanetes, the developmental pattern is different. *Micromonospora* forms single spores, which are not enclosed in a sporangial wall, on the tips of hyphae within the substrate mycelium (× 2,300).

90 *Ancyclobacter aquaticus* (*Microcyclus aquaticus*). The translucent appearance of the colonies is indicative of a low incidence of gas vacuoles. Where gas vacuoles are abundant, the colonies have a chalky white, opaque appearance. (Source: river water; Enrichment: 100 ml water supplemented with 10 mg peptone, incubated at room temperature for two to three weeks; Medium: half-strength nutrient agar + 0.1% peptone; Incubation: 30°C (86°F) , 7 days.)

Species of *Aeromonas* (**91**) are present in highly variable numbers that may correspond closely to the trophic state of the water, while *Chromobacterium fluviatile* (**92**) appears to have a highly variable distribution, even over a small geographical area of the same water system.

A wide range of other bacteria may be found in running water but are more common in ponds and lakes. These include *Cytophaga* and other gliding bacteria, *Serpens* and 'non-motile, Gram-negative, curved bacteria' including *Flectobacillus*. A bacterium of the same type, *Spirosoma*, was, however, isolated from a canalized river in close proximity to a similar bacterium that had colonized lock gates at the water level (**93–95**).

A number of prosthecate bacteria can be isolated

91 A species of *Aeromonas*. The isolate could not be identified with any existing species of *Aeromonas*. This is not an uncommon phenomenon with environmental genera, some members of which are pathogens, and results from the greatest attention being given, quite naturally, to those isolates which are associated with disease. (Source: side channel, canal; Medium: *Aeromonas* medium (Oxoid); Incubation: 28°C (82.4°F), 24 hours.)

92 *Chromobacterium fluviatile*. Isolation could be made consistently from one area of an inland river navigation but not from other areas. Other colonies on the primary isolation plate were common waterborne bacteria. (Source: canal; Medium: quarter-strength nutrient agar; Incubation: 22°C (71.6°F), 3 days.)

93 *Spirosoma linguale* isolated from freshwater by repeated streaking on MS medium. Presumptive identification may be made on the basis of phase contrast microscopy. (Source: bypass weir, canal; Medium: MS medium; Incubation: room temperature, 10 days.)

94 A slime-producing, *Spirosoma*-like organism. Isolation was made from a layer of slime just above the water level of lock gates. Despite the different colonial appearance to typical strains of *Spirosoma*, the cellular morphology, biochemical features and enzyme profiles (API-zym®) were very similar. *Spirosoma*-like organisms have also been isolated from gelatinous deposits on wet planks in a mine. The slime may have a role in protection from dehydration. (Source: slime, lock gates; Medium: MS medium; Incubation: room temperature, 10 days.)

from freshwater. These bacteria are usually present in truly oligotrophic environments, where nutrient levels are very low. In genera, such as *Hyphomicrobium*, the prosthecae serve a reproductive function, but in others the significance is less clear. In bacteria with long appendages, such as *Ancalomicrobium* (**96**) it may be postulated that the prosthecae impart a selective advantage by increasing the efficiency of nutrient uptake by increasing the surface area of the cell and retarding sedimentation.

In some cases, such as *Caulobacter*, the length of the prosthecae increases in response to low nutrient levels. A further feature of some prosthecate bacteria is the possession of a holdfast, usually situated at the tip of the prostheca, or the tip of the cell body. It is thought probable that bacteria with holdfasts, such as *Caulobacter*, preferentially attach to algae, protozoa, cyanobacteria and larger bacteria (**97**) and obtain nutritional benefit from organic secretions from the larger cell. Attachment to inanimate surfaces may also be advantageous nutritionally.

Sheathed bacteria are also a feature of freshwater environments and are often attached to solid substrates by a holdfast. *Leptothrix* is usually found

95 Lock gates on a canal. A thick slime develops at the waterline on the gates and other wooden canal structures. A mixture of Gram-negative bacteria is present, dominated by the *Spirosoma*-like organism.

96 Cells of the prosthecate bacterium *Ancalomicrobium* (× 2,500).

97 *Caulobacter* attached to cells of a larger bacterium (× 4,000).

in unpolluted waters containing metal salts and accumulates ferric and manganic oxides in the sheath. *Sphaerotilus* (98) also accumulates metal oxides but is found in slow-running streams and ditches heavily contaminated with sewage or other organic matter.

Gallionella (99–101) is another structurally distinctive bacterium which, like *Sphaerotilus* may be referred to as an 'iron bacterium'. These bacteria, together with *Leptothrix* and *Sphaerotilus*, promote the corrosion of cast iron and mild steel pipes as well as of austenic stainless steel by the formation of ferric

98 *Sphaerotilus*, showing the sheath enclosing a chain of cells.

99 Unstained smear of *Gallionella* taken from iron- and manganese-bearing water from a stream flowing from an abandoned mine shaft. Photographed using light microscopy.

100 The same growth of *Gallionella* as in **99** but photographed using Nomarski interference microscopy. The spiral stalk of *Gallionella* is readily visible (arrowed).

101 When grown in liquid medium containing ferrous sulphide, *Gallionella* forms 'cottony' colonies attached to the sides of the vessel. These are very distinctive and an effective means of presumptive recognition of the bacterium. **99–101** reproduced with permission from Verran *et al*., 1995, © 1995 Society for Applied Bacteriology.

hydroxide tubercles (Stott, 1988). *Gallionella* can also cause severe fouling of water wells but, as has been noted earlier, nothing is simple in the microbial world and the organism may also be used to remove ferrous iron from shallow aquifer supplies, thus permitting use for domestic purposes (Malkii, 1988).

Primary production in freshwater systems is photosynthesis by higher plants, eukaryotic algae and cyanobacteria, with higher plants being of much greater overall significance than in seawater. The majority of freshwater bacteria are heterotrophic and involved in the degradation of organic matter.

Moulds, especially the aquatic *Phycomycetes* (water moulds) (**102, 103**), are early colonizers of plant and animal materials and are involved in the degradation of complex molecules. In northern temperate waters, leaf material is an important energy source and the hyphomycete *Anguillospora longissima* is of considerable importance in the incorporation of leaf material into the detrital food web (Bermingham *et al.*, 1995). In many running waters the rate of organic turnover is high and mineralization takes place in the water column. Bacterial populations are primarily limited by protozoan predation and

102 Aquatic phycomycetes. (Source: woodland stream containing a high level of fallen twigs and branches; Medium: nutrient agar; Incubation: 20 °C (68°F), 5 days.)

103 Life cycle of chytrids. Chytrids, which include the aquatic phycomycetes, produce motile spores and are the fungi most closely resembling the protozoa. The flagellated zoospore settles on a solid surface and development commences with the production of a branching system of rhizoids. Subsequent growth produces a spherical zoosporangium that cleaves internally to produce zoospores. These are finally liberated by the rupture of the zoosporangium to repeat the cycle.

104 Confluent plaques on an isolation plate. The plaques were the result of cell lysis by presumptive *Bdellovibrio bacteriovorus*. Protozoa and myxobacteria were eliminated by phase contrast microscopy and the delayed development of plaques, together with increasing size during continuing incubation, eliminated bacteriophage. (Source: lowland river, about 250 m downstream of sewage works discharge; Medium: quarter-strength nutrient agar: Incubation: 25°C (77°F), 7 days.)

bacteriophage activity but the parasitic *Bdellovibrio bacteriovorus* can also be present (**104–106**). The relative importance of *Bd. bacteriovorus* in predation is a matter of debate but in overall terms it has been considered to be very limited. *Bd. bacteriovorus*, however, requires a high prey concentration ($c.10^6$ cells ml^{-1}) for survival and it is likely to be concentrated in small ecosystems where its importance may be very much greater. Although gliding myxobacteria are primarily soilborne, the nonfruiting algicidal genus *Lysobacter* (**107**) is present in slowly running water containing large quantities of plant debris (**108**). In some circumstances *Lysobacter* may limit the extent of algal and cyanobacterial blooms.

The extent of deposition of organic matter into sediment varies according to the nutrient load of the water system. In some cases, turbulent flow produces sufficient scouring to maintain most of the sediment in suspension. Particles are subject to bacterial degradation but may be sufficiently large

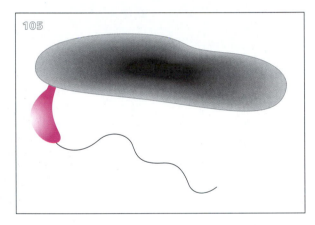

105 Diagrammatic representation of attachment of *Bdellovibrio bacteriovorus* to a prey bacterium. The highly motile *Bdellovibrio* appears to locate its prey by chance. Strike is forcible and rapidly followed by penetration.

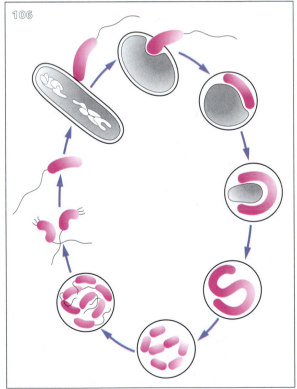

106 Life cycle of *Bdellovibrio bacteriovorus*. *Bdellovibrio* adheres to the prey cell wall by its non-flagellate pole. As it penetrates the cell wall and settles in the periplasmic space, the prey cell becomes rounded, resembling a spheroplast. The *Bdellovibrio* cell continues to elongate until the nutrients derived from the shrinking prey protoplast are exhausted. *Bdellovibrio* then undergoes multiple division to uniformly sized cells, followed by lysis of the prey cell wall and release of free-living progeny. The parasitic life of *Bdellovibrio* is considered to be an extreme example of adaptation to environments of low nutrient concentration and offers a striking contrast with the more generally used strategies.

Feature 6. A lost wilderness

It is well known that the juxtaposition of large human populations with sensitive ecosystems leads to a high level of anthropogenic stress. The Everglades of Florida provide an example of anthropogenic stress affecting life of all types. Residents of, and visitors to, the Everglades must be aware of the commercial development of this wetland ecosystem, especially the all too visible out-of-town shopping malls. The micro-environment is adversely affected by nutrient enrichment, especially of phosphates, from intensive agriculture, altered hydrology as a result of flood control and water abstraction and relatively high mercury levels in the biomass. Sediments emit significant quantities of the 'greenhouse' gases methane and nitrous oxide. This may be a consequence of the nutrient level. In the Everglades, sediments close to agricultural runoff are characterized by low redox potentials, high phosphate concentrations and high water column conductivity. The number of anaerobic micro-organisms is 10^3 to 10^4 times greater than that from sites remote from agricultural runoff. This is probably a consequence of the high speed of response by anaerobes to an influx of nutrients such as phosphates, sulphates and nitrates. Data from Drake *et al.*, 1996.

to support the growth of aerobes, micro-aerophils and anaerobes along an oxygen gradient. Water courses carrying a high load of organic matter may deposit considerable quantities of sediment, especially where slow flowing. Such sediments may be anoxic relatively close to the sediment surface and, depending on sulphate concentration, either sulphate-reducing bacteria or methanogens may be active. Such conditions, however, are more typical of some types of lake (*Feature 6*).

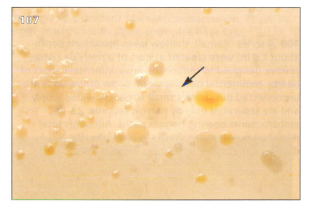

107 A species of *Lysobacter*. *Lysobacter* (arrowed) is one of the more common gliding bacteria in water. A zone of clearing in the yeast cell substrate may be seen surrounding the colony. The source is particularly rich in *Lysobacter* and other gliding bacteria. (Source: inlet off lowland river; Medium: yeast cell agar; Incubation: room temperature, 3 days.)

108 In running water, *Lysobacter* appears concentrated in discrete areas. Several gliding bacteria have been isolated from the inlet illustrated, which is shallow, slow flowing and contains large quantities of plant debris.

114 Snow cover at Estany Rado during winter.

115 The flagellate *Dinobryon cylindricum*, a photosynthetic eukaryote isolated from the Estany Rado snow cover.

116 The ciliate *Perigostrombidium fallax*, a predator on bacteria in the deeper layers of snow cover during the spring melt.

Feature 7. Rocky mountain high

Colonization of sea ice is a well-known phenomenon (see page 55) but it is often not appreciated that active microbial communities can develop in the ice and snow cover of high mountain lakes. Studies in Estany Rado, Spain (114), showed that the winter cover provides a highly dynamic environment, the physical structure and chemical characteristics of which change dramatically as a result of snowfall, periodic melting, freezing and flooding. The main groups of micro-organisms, autotrophic and heterotrophic flagellates (115), ciliates (116) and bacteria (117), respond directly to these physicochemical changes with resulting effects on biomass and species composition. There appear to be two distinct phases which have been attributed to different mechanisms of colonization and growth. From January to mid-April (the phase of growth of the cover), microbial assemblages are related to their planktonic counterparts. During this period, colonization is primarily due to plankton derived from the lake water which floods the cover. As light fails to reach the slush layers, the plankton is primarily dependent on bacterivory for growth.

From mid-April to the spring melt in June (ablation), the physicochemical nature of the environment is strongly influenced by the large amount of water (118) from the melting of the snowpack. The water is a source of both nutrients and micro-organisms, while the light availability permits the growth of algae and an associated food web. In deeper layers the food web is based on bacteria, providing prey for large ciliates, which are probably derived from littoral or watershed soils. 114–116 and 118 reproduced by courtesy of Dr M. Felip, University of Barcelona, Spain; 117 reproduced with permission from Felip *et al.*, 1995, ©1995 American Society for Microbiology.

117 Bacteria, viewed by epifluorescent microscopy after staining with DAPI, from the deeper layers of snow cover during the spring melt. Note the presence of elongated cells; cf. **110**, **111**.

118 Water from melting snow cover at Estany Rado during spring (May). The scale of the melt is apparent from the quantity of water in the foreground.

119 *Legionella pneumophila*. Methods for isolation from natural environments are often time consuming, the most suitable method varying according to the source of the sample (Kusnetsov *et al.*, 1994). Colonies on the non-selective medium illustrated may also have a ground glass appearance. (Source: thermally polluted, disused canal; Medium: buffered charcoal–yeast extract agar supplemented with ∝-ketoglutarate – Edelstein, 1981; Incubation: 35°C (95°F), 7 days.)

120 *Legionella pneumophila*. Cells are highly pleomorphic, varying from short cocco-bacilli to long filaments, though filaments are not present during growth in animal tissue. Staining is often weak and may be irregular. (Source: as for **119**; Microscopy: Gram stain, × 1,200.)

nutrient rich. Actinomycetes are important in the decomposition of organic compounds such as chitin and cellulose. The dominant genus is often *Micromonospora* with *Streptomyces* second in importance (Jiang and Xu, 1996).

Species of the human pathogen *Legionella* (**119**, **120**) are common in natural waters, especially those of high nutrient content and thermally polluted. Waters of this type appear to serve as a reservoir of *Legionella* from which the bacterium enters and colonizes water cooling systems, humidifiers, and the like. Some of these provide a very favourable environment for *Legionella* in terms of temperature, level of aeration, organic nutrients and iron concentration. In either natural or artificial systems the organism may develop as a member of a biofilm but there is also an association between *L. pneumophila* and the free-living amoeba

Relationship between freshwater micro-organisms and aquatic plants

Aquatic plants are of much greater importance in freshwater than in marine environments. The general relationship reflects that of micro-organisms with terrestrial plants. Colonization of roots and their environs is analogous to the rhizosphere and bacteria present can be either beneficial or deleterious to the plant. A similar, but usually less diverse, microflora develops on the roots of plants grown in soilless nutrient solutions.

Colonization of the leaves and shoots of aquatic plants appears variable. In some cases a biofilm develops, members of which are similar to micro-organisms in the water column and colonizing inert material. In other cases, the epiphytic microflora is distinct from that in surrounding water. Reed beds fringing an artificial lake (135) show a distinctive pattern, in which submerged stems of live plants are colonized by a single dominant Gram-negative bacterium (136). This microflora differs from the highly varied microflora in the water of the reed bed (137) and from the water in the body of the lake (138). The epithytic microflora on reed stems above the water level was also distinct from that below (139, 140).

The floating aquatic fern *Azolla* forms an economically important symbiotic relationship with the heterocystous cyanobacterium *Anabaena azollae* (141). *Anabaena* is harboured within mucilage-containing cavities on the leaves of *Azolla* (142). This symbiosis is of very considerable importance in supplying nitrogen to rice paddies in southeast Asia, *Azolla* being co-cultivated with rice. Dense growth of *Azolla* is also beneficial in reducing weed growth and is harvested for use as green manure or poultry feed. *Anabaena* is commercially available as an inoculant and the use of the *Azolla* symbiosis as means of enhancing rice cultivation has been extended to other growing areas.

135 Reed beds, planted during the creation of a nature reserve. Reeds carry oxygen to the roots creating a complex system of microhabitats housing a large microbial population.

136 A Gram-negative, oxidase-positive, rod-shaped bacterium dominant on the submerged surface of reeds. (Medium: quarter-strength nutrient agar; Incubation: 22°C (71.6°F), 5 days.)

137 Microflora of water in a reed bed. A range of Gram-negative rods (oxidase positive and oxidase negative) and smaller numbers of Gram-positive cocci were dominant. (Medium: quarter-strength nutrient agar; Incubation: 22°C (71.6°F), 5 days.)

138 Microflora of lake water near reed beds. Gram-negative rods, including yellow-pigmented and swarming strains, are dominant. (Medium: quarter-strength nutrient agar; Incubation: 22°C (71.6°F), 5 days.)

139 A species of *Erwinia*, one of the two dominant epithytic bacteria of reed stems above the water level. (Medium: quarter-strength nutrient agar; Incubation: 22°C (71.6°F), 5 days.)

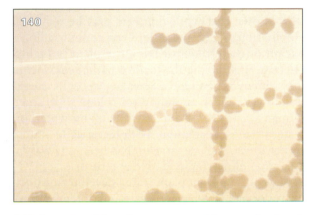

140 A species of *Pseudomonas*, one of the two dominant epithytic bacteria of reed stems above the water level. (Medium: quarter-strength nutrient agar; Incubation: 22°C (71.6°F), 5 days.)

141 Diagrammatic representation of filaments of *Anabaena* (x 700).

142 The *Azolla* symbiosis. In the immature leaf, mucilage-containing cavities are open, but close with maturity. These cavities harbour the filaments of *Anabaena*.

Treatment of sewage and drinking water

Sewage treatment

Mineralization of organic matter in natural waters plays an important role in the cycles of matter. The growth of the human population and its concentration in relatively small geographical areas means that natural processes are no longer adequate to dispose of waste material. One of the prime roles of sewage collection (**143**) and treatment is the intensive operation of substantial parts of the cycles of matter under relatively controlled conditions. A variety of approaches may be taken to sewage treatment but the principle is the same in each case. Solids which form sediment relatively readily are removed by various grit and sedimentation chambers. The liquid then undergoes biological purification that mimics mineralization in the natural environment. For many years the most widely used means was the percolating filter. The most common form involved rotating arms spraying the liquid onto a bed of loosely packed rocks, or clinker, through which it percolates by gravity. Percolating filters are still widely used but have been replaced in many plants by equipment using active aeration. In the activated sludge system, air is blown upwards through a tank of liquid effluent (**144**) while variants include 'racetracks' comprising a system of circular channels through which effluent is pumped under conditions designed to promote aeration. In percolating filters the degradation of organic matter is carried out by communities of micro-organisms in biofilms which develop on the surfaces of the filter, while in activated sludge plants and similar systems, flocs are formed which are composed of actively metabolizing bacteria. The role of micro-organisms is similar in each case and while the process is primarily aerobic, chemical gradients across both the film and floc are such that anaerobic activity also takes place. Treated liquid is discharged into rivers or the ocean.

Sediment removed at the preliminary screening stage, together with sedimented flocs produced by activated sludge processes are passed to anaerobic digestion chambers. Organic input provides sufficient H_2 for both sulphate-reducing bacteria and methanogens. Methane production is very considerable and is commonly collected, compressed and used as a fuel for electrical generators. The sludge remaining after anaerobic digestion is dried, composted and used as fertilizer, or burnt. In the former case there has been concern over the possible survival of pathogens and heavy metal levels.

Although well designed and correctly managed sewage disposal plants produce a highly mineralized liquid effluent, products including nitrate, phosphate, and ammonium are present and can be significant contributors to eutrophication. Further chemical treatments are available to remove these ions or, alternatively, the nitrogen may be removed by denitrification following application to soil or used for fertilization of forests.

For many years it was thought that sewage treatment effectively removed enteric pathogens but it is now known that while numbers may be reduced, elimination is not possible, even though chlorination of the effluent leaving the plant is highly effective in reducing the load of pathogens, including enteroviruses (**145**). Sewage disposal systems do, however, play an important role in separating human waste from sources of drinking water, or water used for recreation. Both historically and in recent times, such separation has been a factor in reduction of morbidity due to enteric disease.

Sewage treatment plants often develop a distinct microflora, in which prosthecate and/or filamentous bacteria are commonly favoured. An example is that of *Hyphomicrobium*, a Gram-negative methylotrophic bacterium that reproduces by budding at the tips of the cellular prosthecae, the hyphae (**146**). *Hyphomicrobium* is one of a group of morphologically related bacteria including *Hirschia*, *Hyphomonas*, *Pedomicrobium* and *Rhodomicrobium*, but differs physiologically in its ability to utilize Cl compounds. *Hyphomicrobium* is found in a diversity of aquatic habitats as well as in soil and its ecological importance may well be under-estimated. The organism attaches to flocs formed in the activated sludge sewage treatment process (**147**) and is notable for its genetic diversity which leads to different, and distinct, populations of *Hyphomicrobium* being present in different parts of sewage treatment plants (Holm *et al.*, 1996).

In the UK, however, there is no statutory requirement to remove pathogens. The basis of sewage treatment is the reduction of Biological Oxygen Demand (BOD) to satisfactory, legally defined, levels.

143 Efficient sewage disposal is taken very much for granted in the western world and municipal engineering is hardly a fashionable subject area. In earlier times, however, the water supply and sewage disposal systems were justifiably viewed alongside railways as triumphs of engineering science and a considerable source of civic pride. Sir Joseph Bazalgette was Engineer to the Metropolitan Board of Works and responsible for the London, UK, sewage and drainage system.

144 An activated sludge plant. Success of the process depends on the microbial community forming flocs, which rapidly settle leaving clear effluent. The process is disrupted by 'bulking', when filamentous organisms extend from the floc into the bulk solution interfering with clarification. Few of the bulking organisms have been isolated in pure culture and the classification of some, such as *Michothrix parvicella*, has been doubted.

146 *Hyphomicrobium* sp. from a sewage treatment plant, showing helically twisted hypha.

145 Poliovirus from raw sewage forming plaques in monkey kidney cell monolayer.

147 An activated sludge floc with attached cells of *Hyphomicrobium*. **146** and **147** reproduced with permission from Holm *et al.*, 1996, © 1996 American Society for Microbiology.

pathogens and the bacterial pathogens *Agrobacterium tumefaciens* (**215**, **216**), the causative organism of crown gall disease (**217**) and *Ag. rhizogenes*, the causative organism of hairy root disease. *Agrobacterium* resembles *Rhizobium* (see page 110) in terms of morphology and physiology and there are similarities in the method of infection by the two micro-organisms (**218**). In many ways a comparison of *Agrobacterium* and *Rhizobium* illustrates the narrow division between parasitism and mutualism. *Agrobacterium* is distinguished from other plant pathogens in that crown gall tumours are capable of growing continuously in the absence of the initiator. This is due to a stable change in cell heredity, genes from the bacterial plasmids being transferred to, and integrated into, the host plant cells. There has been considerable success in exploiting this property and using *Agrobacterium* as vector in the production of transgenic plants (Evans, 1993).

The role of antibiotics in natural environments is always a matter of debate, some microbiologists considering production to be of no consequence, while others consider it to be a major factor in determining the outcome of competition. Broad-spectrum antibiotic production is thought to be a major factor in the antagonistic properties of *B. subtilis*, an important potential biological control agent against fungal pathogens. It seems likely that volatile compounds are also involved (**219–224**).

215 *Agrobacterium tumefaciens*. Colonies are unremarkable and either non-pigmented or a light beige-brown. Copious slime is produced on carbohydrate-containing media. (Source: soil around infected seedlings; Medium: nutrient agar; Incubation: 25°C (77°F), 2 days.)

216 *Agrobacterium tumefaciens*. Aggregates which are distinctly star or rosette shaped may be seen but this property appears to be strain dependent. Much longer cells are common in older cultures. (Source: as for **215**; Microscopy: Gram stain, × 1,200.)

217 Tumour caused by *Agrobacterium tumefaciens* on a young tomato plant. Reproduced by courtesy of Prof. D.G. Jones, Department of Agricultural Sciences, University of Wales, Aberystwyth, UK.

218

A. *Rhizobium*
1. Gram-negative rod-shaped bacterium.
2. Free-living cells of *Rhizobium* are stimulated by proximity to roots of host (leguminous) plants.
3. Infection thread enters root hairs, may be through wound.
4. Tetraploid host cell followed by diploid host cells stimulated to divide repeatedly to form nodule.
5. *Rhizobium* in nodule loses ability to divide, but maintains metabolic activity for extended period.

B. *Agrobacterium*
1. Gram-negative rod-shaped bacterium.
2. Free-living cells of *Agrobacterium* are stimulated by roots of host.
3. Enters root hair *via* wound (may also enter other parts of plant).
4. Host cells transformed by *T-DNA* from plasmid to stimulate cell division and formation of tumour.
5. After initial period when growth of *Agrobacterium* stimulated by host-produced opines, bacterium loses ability to multiply while remaining metabolically highly active.

218 Comparison of infection with *Agrobacterium* and *Rhizobium*.

219–224 Effect of *Bacillus* volatiles and antibiotics on the pathogen *Rhizoctonia solani*. In comparison with control hyphae (**219**), volatiles induce vacuolation (**220**) and swelling of the hyphae tip (**221**). Swelling of hyphae tips also occurs in the presence of *B. subtilis* antibiotic (**222**) accompanied by cytoplasmic breakdown within the hyphae (**223**) and lysis of the tips (**224**). **219–224** reproduced with permission from Fiddaman, P.J. and Rossall, S., 1993, © 1993, The Society for Applied Bacteriology.

Mycorrhizal relationships

Mycorrhizal relationships have existed since the Devonian period and some primitive plant orders, such as the Magnoliales, have an obligate dependence on mycorrhizae. Some flowering plant families, however, contain non-mycorrhizal species.

At its simplest, the mycorrhizal relationship is seen as a straightforward mutualistic relationship. The fungi that participate are usually those unable to utilize the complex polysaccharides, which are the principal carbon sources for micro-organisms in many types of soil. Mycorrhizal associations enable these fungi to obtain an abundant supply of simple carbohydrates, such as glucose, from the plant. Fungi of mycorrhizal species may be free-living in soil but generally they have only a limited ability to exist saprophytically.

Infection of plants with mycorrhizal fungi involves a complex interaction involving plant, soil, fungi and other soil micro-organisms. Soil extracts, host root exudate and, specifically, flavonoid compounds are all stimulatory to vesicular–arbuscular (VA) fungi but knowledge of the communication between roots and fungi is incomplete.

The plant benefits from access to nutrients which are otherwise unavailable. This results from the ability of mycorrhizal fungi to hydrolyse large nitrogen- and phosphorus-containing molecules. In addition, hyphae, or an aggregation of hyphae, and fungal cords widen the area of root exploration and are able to bridge zones of low nutrient concentration. Cords are also involved in water transport to the root, thus enhancing drought resistance. Root size and longevity can also be increased by vitamin production by mycorrhizal fungi.

Two broad categories of mycorrhizae exist: ectotrophic and endotrophic. In addition, a small group of ecto/endotrophic mycorrhizae exist, variously termed arbutoid or monotropoid depending on the type of plant partner. Endotrophic mycorrhizae are the most widespread and are further categorized into vesicular–arbuscular (VA), ericoid and orchidaceous. Of these, VA mycorrhizae are most common and, in many cases, are associated with the roots of economically important plants. Endotrophic mycorrhizae are found on about 90% of the world plant flora, though ectotrophic mycorrhizae (**225, 226**) are of considerable importance in temperate forests. At least 70 fungal species are known to be capable of forming mycorrhizal associations and the actual number is undoubtedly many times greater. With a few exceptions, mycorrhizal fungi are not species

225 An ectomycorrhizal root system developing on the root system of a pine seedling.

226 Fungi forming ectomycorrhizal associations are mainly members of the Agaricales, Hymenomycetes and Basidiomycotina. They form conspicuous fruiting bodies (such as mushrooms and toadstools) on forest floors in autumn and spring.

specific. A given fungus can be associated with several plant hosts while equally a given myco-trophic plant can form associations with one of several fungi. One species of pine, for example, can form associations with any of more than 40 species of mycorrhizal fungi. Despite this lack of specificity, a predominant mycorrhizal type is recognized in each different ecosystem. This is a consequence of Liebig's 'Law of the Minimum' which states that, when all other nutrients are available, the growth of plants is regulated by supplies of that nutrient element that is present in the smallest amount. This 'law' may be extended to heterotrophs such as fungi. Plants growing in different ecosystems face different nutrient limitations, selection favouring the fungi that can most effectively mobilize or capture the limiting nutrient for plant growth in that ecosystem (Read, 1991).

In heathland, and in most boreal and temperate forests, nitrogen is the 'minimal' nutrient and mycorrhizal fungi are able to mobilize that element. Further selection occurs according to the nature of the nitrogen-containing resources. In heathland, nutrients are sequestered in an acidic organic matrix ('mor' humus) that consists of highly recalcitrant residues of the ericaceous plants (e.g. heather) and fungi which are able to grow in this ecosystem. Fungi participating in ericoid mycorrhizae are thus of physiological types capable of hydrolysing lignin, proteins, polyphenols and chitin. In the less severe environments of the boreal and temperate forests, substrates are intermediate in terms of recalcitrance. Plants are of the ectomycorrhizal type, the hydro-lytic activity of the participating fungi varying according to the substrate, and being greatest where residues are of high C:N ratio. Only a few ecto-mycorrhizal fungi are able to break down very complex molecules such chitin but, in general, this group has wider hydrolytic properties than realized previously.

In warmer climates, grasslands and tropical forests develop, the substrates in these exosystems being the least recalcitrant. Phosphorus, rather than nitrogen, is the growth-limiting nutrient and VA mycorrhizae are dominant. In the great majority of cases, VA fungi are of very limited hydrolytic ability and largely unable to mobilize minerals from polymers. However, an extensive hyphal network develops that is highly effective in scavenging soluble phosphate ions. It is also possible that VA fungi are directly involved in the mobilization of phosphorus from recalcitrant sources (Gianinazzi-Pearson and Gianinazzi, 1989).

Although the nutritional aspects of mycorrhizal associations are often considered of greatest importance, the plant may gain other significant advantages. Mycorrhizae may protect the root from pathogens by acting as a mechanical barrier or by secreting antibiotics. Indirect protection may also occur by alterations to the rhizosphere microflora and stimulating competition for ecological niches occupied by pathogens. Mycorrhizal fungi are also of importance in protecting host plants from the toxic effect of pollutants, including acid rain, sulphur dioxide and metal ions, especially heavy metals, aluminium and, at high concentrations, iron. Metal toxicity can be a major problem in soils with low pH value as a consequence of the greater solubility of the metallic ions; ericoid mycorrhizae are often of particular significance. The situation with iron is complicated by the fact that the metal is an essential element at lower concentrations. Under these conditions, ericoid mycorrhizal fungi enhance plant growth by significantly increasing iron uptake. A number of species have been shown to produce hydroxamate siderophores which are probably involved in scavenging for iron when the metal is present only at very low concentrations (Schuler and Haselwandter, 1988).

Mycorrhizal infection is of economic and com-mercial significance due to the importance in the establishment and growth of plants. In some situ-ations inoculation of plants with the mycorrhizal symbiont is considered necessary (*Table 3*).

Table 3. Applications of mycorrhizal inoculation.

Orchid culture*
Introduction of exotic trees to new sites
Introduction of ericaceous fruit crops
  (e.g. blueberry) to new sites
Propagation of woody ornamentals after soil
  sterilization
Planting of Sitka spruce on first-rotation forest
  sites
Establishment of trees during remediation of
  virgin sites, such as coal spoil or waste tips

*Inoculation not now usual commercial practice
Data from Mitchell (1993)

Nitrogen-fixing symbioses

The importance of leguminous plants in crop fertility has been recognized since antiquity. This is a consequence of infection of the plants with nitrogen-fixing rhizobia (*Rhizobium* and *Bradyrhizobium*). The rhizobia are present in variable numbers in normal soil but inoculation of the seed of legumes with pure cultures of rhizobia is a well-established practice to ensure high yield. Alternatively the inoculant may be sprayed into the soil, while dual inoculation with rhizobia and VA mycorrhiza has also attracted attention. Leguminous plants stimulate the growth of rhizobia in the soil to a very considerable extent, effects extending as far as 25 mm from the root surface. Stimulation of free-living cells is highly specific, only *Rhizobium* (**227**) and *Bradyrhizobium* benefiting.

Both *Rhizobium* and *Bradyrhizobium* are highly host-specific. *Bradyrhizobium* is unique in forming a symbiotic relationship with the non-legume *Parasponia*. It has been postulated that *Bradyrhizobium* (which is alkaline-producing and highly suited to life in acid, tropical soils) is the ancestral form from which the acid-producing *Rhizobium* eventually arose, but this is not universally accepted.

Infection of the host plant (**228**) is initiated by penetration of the root hair by a group of *Rhizobium* cells. Invagination of the root hair membrane leads to formation of an infection thread which penetrates the root cortex. Bacterial cells are ultimately liberated into the host cell cytoplasm enclosed in a membrane. The characteristic nodule (**229**, **230**) is formed when the infection thread reaches a tetraploid cell of the cortex which, along with neighbouring diploid cells, is stimulated to constant division. Young bacterial cells in the nodules are rod-shaped but later become irregular in shape as metabolically active but non-dividing 'bacteroids'. Nitrogen-fixation continues so

227 *Rhizobium leguminosarum*. Isolation from nodules is straightforward, involving washing, sterilizing and rinsing the nodule. The nodule is then crushed in a small amount of sterile water and the resulting suspension streaked on to solid medium. (Source: nodules, white clover; Medium: yeast extract–mineral salts–mannitol; Incubation: 28°C (82.4°F), 3 days.)

10 mm

228 Infection of leguminous plants with *Rhizobium*. A diagram of a newly infected root hair shows the bacterial infection thread passing up the hair. Infection results in a progressive curling of the root hair tip.

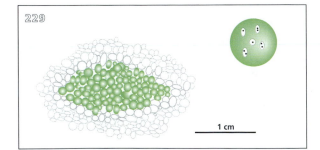

1 cm

229 A section through a characteristic nodule illustrated diagrammatically. The green cells indicate those filled with bacteria. The magnified part shows one of the mature nodule cells having a large membrane envelope containing four to six 'bacteroids'. No further bacterial growth occurs.

long as the cells are metabolically active. Free-living cells do not fix N$_2$.

Root nodules are also produced in a nitrogen-fixing association between the actinomycete *Frankia* (**231**) and a very wide range of non-legumes. Nodules are of two types, negatively geotrophic roots in the Casuarinales and Myricales and repeatedly branching roots in other species. The role of cyanobacteria in nitrogen-fixing symbioses is also important although root nodules are not formed (**232**).

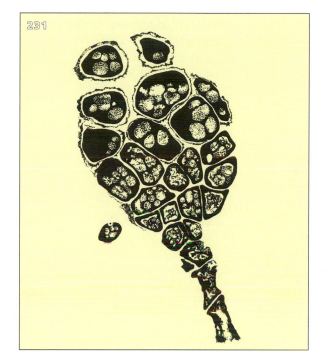

230 *Rhizobium* nodules (arrowed) on the roots of young, white clover. Reproduced by courtesy of Prof. D.G. Jones, Department of Agricultural Sciences, University of Wales, Aberystwyth, UK.

231 Section through a sporangium of *Frankia*. Mature spores are at the apex, young spores developing at the base of the sporangium (x 11,500).

232 The first stage in symbiosis between the cyanobacterium *Nostoc* and liverwort (*Blasia*). Reproduced by courtesy of S. Babic and D.G. Adams, University of Leeds and the Society for General Microbiology.

Micro-organisms on the surfaces of plants

The aerial surfaces of plants are colonized by a mixed and characteristic microflora. This is commonly referred to as the epiphytic microflora although this is not an absolute classification since surface micro-organisms may become internal following heavy rain and possibly under other circumstances (Hirano and Upper, 1988). Micro-organisms on plant surfaces as a whole have been placed in two ecological groupings: 'residents' (capable of colonization) and 'casual' (transitory).

Bacteria, yeast and fungi are all present as colonizers of plant surfaces. There is, however, a microbial succession which determines the dominant type. Bacteria are the main early colonists and dominate during the early growing season. Yeasts and yeast-like fungi become of greatest importance during the middle of the growing season and are of overall greatest importance, outnumbering other micro-organisms by up to 50 to 1 (233–242).

233–235 Epiphytic microflora of pampas-grass (**233**), tragescanthia (**234**) and rose (**235**). All plants were grown in the same area and shared some members of the microflora. Pampas-grass, however, was less heavily colonized while only tragescanthia (grown indoors) was colonized by *Streptomyces* (arrowed; blue-grey colonies superficially resembling moulds).

236 The epiphytic microflora of pampas-grass was characterized by slow-growing yeasts and yeast-like fungi which could not be identified with known genera. These are illustrated on a primary isolation plate (see also **237–239**). (Medium: malt extract agar; Incubation: 22°C (71.6°F), 7 days.)

237–239 Slow-growing yeasts and yeast-like fungi from pampas-grass illustrated on a primary isolation plate (**237**) and after initial isolation (**238**, **239**). (Medium: malt extract agar; Incubation: 22°C (71.6°F), 7 days.)

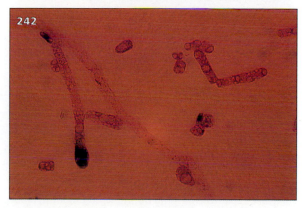

240–242 Photomicrographs of cells of yeasts and yeast-like fungi isolated from pampas-grass. (Microscopy: simple stain; × 40.)

Filamentous fungi follow yeasts and yeast-like fungi (243–247) at the end of the growing season, their hyphae possibly penetrating the tissues and causing minor disease and yield-loss. Micro-organisms may also overwinter in dead plant parts (248, 249).

The epiphytic microflora of plants is pre-dominantly saprophytic although many plant

243–245 Scanning electron micrographs demonstrating interactions between *Trichoderma harzianum* and nylon fibres coated with purified lectin from the plant pathogenic fungus *Sclerotium rolfsii*. The first stage (**243**) involves atypical branching of the *Trichoderma* and contact of the branch tip with the fibre surface (bar = 10 μm). This is followed (**244**) by elongation of the firmly attached tip along the fibre surface (bar = 1 μm). The final stage (**245**) involves tight adherence of the *Trichoderma* hyphae to the fibre surface and the production of additional branches (bar = 10 μm).

246, 247 Scanning electron micrographs of the mycoparasite-related structures formed by *Trichoderma* during interaction with lectin-treated fibres. Appressorium-like bodies (**246**) are formed at the tip of short branches, while hyphal loops (**247**; arrowed) are distinctive features (bars = 10 μm). **243–247** reproduced by courtesy of Dr J. Fabar, Hebrew University of Jerusalem, Israel.

248, 249 Dead plant material provides a large number of microenvironments which can be of importance in the survival of epiphytic bacteria during winter months. Bracken (**248**) supports a relatively wide range of yeasts while pigmented, Gram-negative bacteria could be isolated from teasel heads (**249**) but not the surrounding material.

pathogens, primarily bacteria, can adopt a resident saprophytic phase on either host or non-host plants. Most of the bacteria present are Gram-negative rods (**250**, **251**) although Gram-positive bacteria may be present as 'casuals'. The most common bacteria are *Erwinia* (**252**, **253**), *Pseudomonas* and *Xanthomonas* (**254**) together with an ill-defined group usually described as flavobacteria.

250, **251** Gram-negative, rod-shaped bacteria on the upper (**250**) and lower (**251**) surfaces of barley leaves. Fungal spores and hyphae are also present on the upper surface (bar = 10 μm in **250** and 1 μm in **251**). Reproduced with permission from Blakeman, J.P., 1991, © 1991, Society for Applied Bacteriology.

252 *Erwinia carotovora*. Although most strains of *E. carotovora* are lactose positive, lac⁻ variants are common. Although *E. carotovora* grows well on most common media for the *Enterobacteriaceae*, there is usually no means of differentiation from other members of the family. (Source: infected carrots; Medium: MacConkey purple agar; Incubation: 30°C (86°F), 2 days.)

253 Soft rot lesion on fennel caused by *Erwinia carotovora*. In addition to its role as a plant pathogen, *E. carotovora* is an important cause of spoilage of a wide range of vegetables. The plant may be infected before harvest or the organism may gain entry through post-harvest damage, tissue being subsequently broken down by pectinolytic enzymes. A variety of other micro-organisms may be found in the rotting part but usually have no causal role.

254 *Xanthomonas campestris*. When grown on medium containing high levels of usable carbohydrates, the colonies are mucoid or slimy due to the production of extracellular polysaccharides (xanthan gums). Most, but not all, isolates are yellow-pigmented, the pigments being highly characteristic xanthomonadins (brominated aryl polyenes). Isolation is best made from infected plant material, nutrient agar being satisfactory for most species. (Source: infected pea plants – *Pisum sativum*; Medium: nutrient agar + 5% sucrose; Incubation: 25°C (77°F), 3 days.)

Pathogens with an epiphytic resident phase include the economically important *Erw. amylovora* (fireblight) and *Erw. carotovora atroseptica* (black-leg and tuber rot of potato). The epiphytic micro-flora is also a source of ice nucleation active (INA) bacteria. Specialized pathogens (which act to form nuclei ice crystals) cause, usually localized, damage to plants during frost. The most common yeasts are *Aureobasidium pullulans*, *Cryptococcus* spp. (**255**), *Rhodotorula*, spp. and *Sporobolomyces* spp.

Life on the surfaces of plants is rigorous and provides higher levels of environmental stress than other plant-associated habitats such as roots. Water is usually available only intermittently and cells are subject to damage by UV irradiation. In hot climates, surface temperatures may be sufficiently high to damage more sensitive cells. Micro-organisms are also subject to inhibition by plant-produced inhibitors such as phytoalexins; levels of pollutants may also be high. Nutrients are derived from leachates, originating within the plant. Considerable competition exists between epiphytic micro-organisms for niches on the plant surface. A mechanism known as exclusion is operative in which the first bacterium to arrive at a niche is able to exclude other bacteria, often closely related, with the same site preference (Blakeman, 1991). A number of competitive strategies may be used, including both antibiotic and bacteriocin production. In general, antibiotic production is limited to a small proportion of the epiphytic microflora. *Pseudomonas fluorescens*, however, produces both

antifungal and antibacterial antibiotics and *in vivo* inhibition of the pathogen *Ps. syringae* pv. *phaseolicola* can be demonstrated.

Yeast and yeast-like fungi were not previously thought to produce antibiotics, but production has now been demonstrated (McCormack *et al.*, 1994). Strains of both *Erwinia* and *Pseudomonas* have been shown to produce bacteriocins against closely related bacteria. The *in vivo* significance of bacteriocin production is difficult to evaluate in many situations but there is some evidence of involvement of bacteriocins in the inhibition of *Erw. amylovora* by *Erw. herbicola*.

Despite antibiotic and bacteriocin production by plant epiphytes there is evidence that at least part of the basis for competition is nutritional. This is likely to involve specific nutrients. There is strong nutritional competition between, for example, saprophytic bacteria and necrotrophic fungal pathogens such as *Botrytis cinerea* and *Phoma betae*. The outcome is that amino acids become limiting, with adverse effects on the ability of the fungi to germinate (Blakeman and Brodie, 1977). The exclusion mechanism may also be explained, at least in part, by the enhanced ability of bacteria established in niches to obtain nutrients. Antibiotic production and nutritional limitation may, however, be linked. In pure culture, antibiotic production by yeast was greatest when nutrients were limiting (McCormack *et al.*, 1994). This suggests that antibiotic production is of importance in enhancing the ability to compete for nutrients.

Attempts have been made to exploit the exclusion mechanism to prevent pathogens from becoming established as part of the epiphytic flora. Some success has been achieved in excluding INA bacteria with either mutant INA-negative strains or artificially constructed, isogenic, INA-negative strains. Experimental work is also being conducted in attempts to control *Ps. syringae*.

Some of the epiphytic bacteria are involved in the breakdown of large masses of vegetation after its death. Numbers of *Enterobacteriaceae* increase initially, followed by lactic acid bacteria as the pH value falls and micro-aerophilic conditions develop. Natural heating can mean a rapid temperature rise and the growth of thermophilic bacteria (see 'Landfill', page 122). Large shrubs and trees break down only slowly and provide a wide range of habitats for micro-organisms (**256**) and higher life forms, including small mammals. Micro-organisms involved are primarily fungi (**257**, **258**) although slime moulds (Myxomycetes; **259–262**) and bacteria are also present.

255 A yeast of atypical colonial morphology resembling *Cryptococcus*. The yeast was isolated from beech leaves in several parts of a forest during late summer but not from surrounding soil. (Medium: oxytetracycline– glucose–yeast extract agar; Incubation: 22°C (71.6°F), 5 days.)

256 Colonization of a fallen tree trunk. A tree trunk, which had fallen into a semi-waterlogged area, was colonized by a remarkable variety of micro-organisms including bacteria, microfungi and protozoa. Nematodes were also present in large numbers. No systematic evaluation of bacterial species was made but diversity was considerable. 'Colonies' developing in the slime matrix were polymicrobic aggregations comprising morphologically distinct bacteria and, possibly, yeast.

257, 258 Fruiting bodies of higher fungi are common on decaying wood as well as on live trees. The presence on decaying wood does not necessarily mean that the fungus is of a wood-rotting species.

259–262 Slime moulds (Myxomycetes) are common on decaying wood including wood used for constructional purposes.

The lichen symbiosis

Lichens are composite organisms consisting of a fungal partner (which is usually dominant) and one or more photosynthetic partners (**263**). The composite structure is generally considered to behave as an independent entity. The fungal partner, the mycobiont, is usually an ascomycete, although basidiomycetes are occasionally lichenized. The photosynthetic partner, the photobiont, may either be a green alga or a cyanobacterium. Some lichen symbioses involve three, four or more partners. Although the mycobiont is dominant in terms of providing the bulk of the tissue, the success and longevity of the symbiosis ultimately depends on the photobiont. Photobionts come from a small number of genera, generally widespread in nature, while mycobionts are taxonomically diverse but exclusively lichen forming (*Feature 11*).

From an ecological viewpoint, the lichen symbiosis is considered to be clearly mutualistic since it enables both partners to colonize areas where neither could survive alone. The association may therefore be described as obligate in that the '... fitness of both partners approaches zero when they are not together' (Lewis, 1987). From a purely mechanistic viewpoint, the mycobiont is dependent on the photobiont for organic nutrients. The photobiont is nutritionally independent of the mycobiont but appears to benefit by the enhanced uptake of

water and minerals and protection from desiccation. Parasitic lichens, of the genus *Rhizocarpon*, are also recognized – these are parasitic on other lichens, including other *Rhizocarpon* species (Poelt, 1991).

Feature 11. Life in the fast lane

The lichen symbiosis is one of the most successful known; lichens colonize a vast range of microhabitats in all parts of the world. They are well known as colonizers of common objects such as stonework, rocks, footpaths and glass surfaces (**264–268**). Many are surface dwellers and, as a consequence, can face extreme conditions. The temperature at the surface of a rock, for example, may rise to as high as 70°C (158°F) during the heat of the day but fall below 0°C (32°F) at night. The lichen symbiosis dominates huge areas in the boreal zones and also enables fungi to persist in some of the hottest deserts of the world, in the highest mountain ranges, in tropical rain forests and inside rocks in Antarctica (Hawksworth, 1988). The true importance of the lichen symbiosis is, in fact, restricted to extreme environments and tends to break down during prolonged periods in favourable environments, the photobiont reverting to a free-living form.

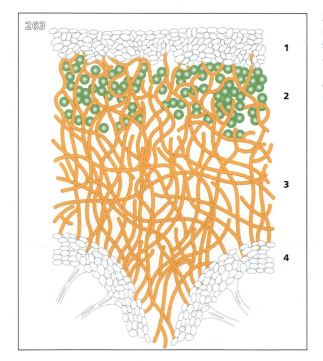

263 General structure of a lichen thallus. Three types are recognized: *crustose* which adhere strongly to the substrate; *foliose* which are leaflike and more loosely attached to the substrate and *fructiose* which form pendulous strands or upright stalks. Note: 1. upper cortical layer; 2. alga (phycobiont); 3. fungal hyphae (mycobiont); 4. lower cortical layer.

264–266 *Et in Arcadia ego.* Lichen-encrusted tombstones are an evocative sight in a country churchyard (**264**). Most on stone in the UK are of the crustose type (**265**). Lichens are highly adapted to their microenvironment and apparently minor changes caused, for example, by a tombstone tilting with age, can upset the balance between the lichen and its surroundings and lead to death (Hawksworth, 1990). Lichens are sensitive to heavy metals and are unable to develop on tombstones where lead lettering is used (**266**).

267, 268 A solitary foliose lichen (**267**) on a fence post representing successful colonization of a microhabitat. A few feet away more favourable conditions on a fallen tree (**268**) permit wider colonization.

From a global viewpoint, lichens are significant in a number of ways. As carbon sinks, lichens fix vast but unquantified amounts of carbon through photosynthesis. Triterpenoids, produced in large quantities by large, rainforest lichens, may contribute to the pool of buried carbon, usually considered to consist of recalcitrant compounds. Such triterpenoids probably have an anti-herbivore function and prevent the utilization of rainforest lichens as a protein source by herbivores. Lichens, however, also have a significant role in nutrient cycling and the food web.

Relatively fast-growing lichens, which have a short life-cycle on short-lived substrates, are of importance in tropical environments. In most cases the growth of lichens is slow and can continue for centuries. Competition is rarely a problem as lichens develop in continuous equilibrium with their immediate environment though the balance is delicate and highly susceptible to disturbance by environmental change. This very susceptibility, however, means that lichens are of considerable value as environmental monitors. Disturbance affects the presence or absence of species and may also involve reduction in frequency, luxuriance, fertility and the ability to establish new colonies (Hawksworth *et al.*, 1974). Lichens are most widely used as biomonitors in studying air pollution and are particularly sensitive to sulphur dioxide (**269**). Lichens have also been used to measure other atmospheric pollutants (Hawksworth, 1990; Richardson, 1991) and, on a much wider scale, climatic change (*Table 4*; Galloway, 1992; Smith, 1990).

Man also makes some direct use of lichens, including medicines (**270**), foods, dyes and as perfume ingredients (*Table 5*). Lichens also interact with man in other ways: although none is a recognized pathogen, secondary metabolites are responsible for contact dermatitis amongst forestry workers (woodcutter's eczema). Allergic responses to perfumes and aftershave lotions are often also associated with the use of lichen extracts (Richardson, 1991).

269 Changes in distribution of *Usnea* in mainland Britain following the implementation of the Clean Air Acts. Solid lines indicate areas where lichens disappeared during 1800–1970. Arrows indicate the advance of recolonization.

Table 4. Use of lichens in environmental monitoring.

Environmental area	Monitoring for
Air pollutants	Sulphur dioxide
	Heavy metals
	Radionuclides
	Fluorides
	Acid rain
Climatic changes	Global warming
	Ozone thinning
	(increased UV radiation)
	Dating of rock surfaces

270 Structure of usnic acid, a lichen secondary metabolite (lichen acid). This compound was previously used as a topical antibiotic.

Table 5. Uses of lichens by man.

Area of use	Use
Medicine	Topical antibiotics such as usnic acid (Usno®)*
	Possible future antibiotics
	Use of Iceland moss *Cetraria islandica* in home-made tonics and laxatives
	Use of *Cetraria islandica* or *Usnea* spp. in commercial remedies for sore throats
Food	Traditional food in times of hardship
	Japanese delicacy prepared from *Umbilicaria esculenta*
	Curry additive Garam masala contains a high proportion of various lichens
Perfumes and toiletries	Use of *Evernia prunastri* (oakmoss) and *Pseudevernia furfuraceae* (treemoss) as fragrance and fixative in aftershaves, perfumes and soaps
Miscellaneous	Use of *Cladonia stellaris* in wreaths, floral decorations and architects' models
	Manufacture of the acid–base indicator litmus from *Roccella* spp.

*Commercial production now ceased due to uncertainty of raw material supply
Information from Richardson (1991)

OTHER TERRESTRIAL HABITATS

Landfill

Municipal waste, which includes domestic rubbish and small, but variable, quantities of industrial and building waste, is usually disposed of in landfill. Landfill sites may also receive various types of clinical waste. The most acceptable approach is the sanitary or engineered site, where waste is covered with soil and compacted immediately after deposition. Leachate is contained and treated and the site is managed to minimize nuisance and avoid pollution.

A very wide variety of materials are present in landfill and a correspondingly wide variety of degradative processes occur. Three major processes are involved: decomposition of cellulose-based materials to sugars, formation of weak acids from the sugars and methanogenesis. Open sites are initially aerobic but sanitary sites become anaerobic very quickly. In either case a temperature rise to $c.60°C$ (140°F) is common and encourages thermophilic micro-organisms including species of *Bacillus* and, at suitable oxygen tensions, the mycelial *Thermoactinomyces*.

Very considerable methanogenesis occurs in landfill. In some cases sanitary landfill sites have been constructed to allow the recovery of methane and its subsequent use as fuel. In many cases, however, methane is a cause of considerable problems, leakage from sites being a nuisance and contributing to the greenhouse effect. Methane seeping from old landfill sites can cause problems in modern buildings erected many years after use of the site has ceased, hence modern site management often involves capping the site with an impermeable clay coat; this procedure is expensive and not always effective. A possible alternative and more economical management strategy is to maintain an open, permeable soil cover and manipulate conditions to optimize the microbial oxidation of methane. In the longer term, however, a better policy is to minimize use of landfill (*Feature 12*).

In the UK, considerable publicity has been given to the finding of clinical material, as well as blood- or faeces-stained material at landfill sites. This has raised fears of environmental contamination with pathogens. Evaluations have shown that the survival of many pathogens in landfill is limited. Some movement in leachate has been reported but, in general, the risk is considered to be low (Collins and Kennedy, 1992).

Lithobiotic (rock-inhabiting) habitats

Lithobiotic micro-organisms are widespread in nature although relatively little is known of their ecology. They colonize those exposed rock surfaces that are not soil covered or occupied by more aggressive higher organisms such as mosses or sessile animals. A wide range of climates and habitats support the growth of lithobionts ranging from both hot and cold deserts to submerged marine and freshwater rocks, and from deep caves to the surface of buildings (Friedmann and Ocampo-Friedmann, 1984).

There is evidence for involvement by micro-organisms with cyanobacteria, acid-producing fungi, thiobacilli and nitrifying bacteria being particular candidates (**271–273**). Testate protozoa (*Difflugia* and *Euglypha* spp.) may also be present. Microbial degradation of the built environment is not, of course, restricted to stone (**274**).

Feature 12. So where does it all go?

Despite the widespread assumption for an anaerobic route of mineralisation in landfill, and the observed quantities of methane evolved, the amount of the gas is usually less than 10% of that theoretically predicted. Recognising and understanding the limitations of methanogenesis in landfill is seen as being of increasing importance. In some cases at least, low rates of methanogenesis can be directly related to the thermodynamic uncoupling of fatty acid oxidation although underlying causes are probably low pH values and an accumulation of toxic, volatile, fatty acids. Data from Mormile *et al.*, 1996.

271–273 Deterioration of building stone. Many factors are involved and chemical pollution is generally thought more significant. It is, however, considered that micro-organisms have a strong influence when cell numbers reach 10^4 to 10^6 g^{-1}. Corroded sandstone (**271**) bore a mixed bacterial and yeast microflora, but the role in corrosion is not known. Adjacent brickwork (**272**) was contaminated with low numbers of a species of *Bacillus*. This organism had a characteristic colonial morphology and was alkali resistant (**273**). The *Bacillus* was probably a secondary invader. The organism could not be isolated from surrounding soil although a morphologically similar *Bacillus* was isolated from 'sarsen stones' at Fyfield Down, Wiltshire, UK. Again, attempts at isolation from surrounding soil failed. (The sarsen stones at Fyfield Down are the remains of a limestone cap which once covered southern England. The down was the source of stones used to build the great stone circles at nearby Avebury and has been described as the finest prehistoric landscape in Europe.)

274 Microbial colonization of interior walls. The growth of mould on interior walls is commonplace and usually associated with condensation or penetrating moisture. In this instance the affected wall showed little sign of mould growth although the paint was badly discoloured and the wall 'sticky' to touch. Culture of swabs taken from the wall showed heavy mixed growth of mould, together with bacteria (1) and yeast (2).

From the viewpoint of man, lithobiotic micro-organisms occupy desolate and hostile environments but it is important to distinguish between conditions in the macroclimate and those immediately surrounding the microbial communities; the term 'nanoclimate' has been proposed for the latter. Under most circumstances rock must be regarded as an extreme environment. Even in the humid north European climate, stone as a biotope offers conditions for organisms similar to those of deserts and semi-deserts (Truper and Galinski, 1986). Lithobiotic micro-organisms are subject to extremes of temperature, desiccation and, at times, of low moisture content, salt stress due to the dissolved salt concentration increasing to crystallization.

Lithobiotic bacteria, and possibly other micro-organisms, possess both an external and an internal cellular protection system. External protection is offered by life in biofilms. The biopolymer matrix itself conserves water through its nature as a hydrogel but a more specific mechanism is the accumulation, in microniches, of hygroscopic metabolites such as $NaNO_3$ which form 'water nests'. Water thus remains available during drought conditions, permitting the survival of bacteria and other micro-organisms sensitive to total dehydration (Moser et al., 1990). Internal protection is offered by synthesis of protective substances such as mono- and disaccharides, amino acids and betaines, which minimize the effect of salt stress (Bock and Sand, 1993).

Three main categories of lithobionts have been proposed although terms are not absolute and many types are best described as intermediate: hypoliths reside under pebbles and small stones on soil (they are restricted to hot desert pavements and are largely cyanobacteria, green algae and heterotrophic bacteria), endoliths live inside rocks, while epiliths live on the surface.

Endolithic micro-organisms are of particular note in their ability to exploit relatively favourable nano-climates within the rock, contrasting markedly with the abiotic interior. Endoliths have been further classified into euendoliths, chasmoendoliths and cryptoendoliths. Euendoliths, which actively penetrate the rock fabric, are absent from extreme arid environments, probably because penetration requires a minimum water content. Euendolithic micro-organisms impinge directly on man in that the exterior surfaces of buildings, statues, and the like, are among the substrates colonized. Constructional stone inevitably undergoes physical and chemical deterioration, the rate of which has been increased by air pollution, especially acid rain. Chasmoendoliths dwell within fissures and cracks, which provide sufficient space for microbial growth. Only certain rock types provide a suitable substrate. These are usually weathered rocks such as granite, calcite, quartzite and dolomite. As with other types of endolithic communities, only light-coloured, or translucent, rock permits sufficient light penetration for photosynthetic micro-organisms which are invariably part of the community. Chasmoendolithic environments permit microbial life in both hot and cold extreme arid areas.

Cryptoendoliths colonize porous rocks, such as sandstone, which have sufficient air space to permit microbial growth. The morphology of colonization is little affected by the macro-environment. The outer 1–3 mm of the rock forms a crust in which no micro-organisms develop. Colonies develop in the next few millimetres, adhering to or growing between crystals. The airspace is thus sealed from the exterior by the outer crust of rock. This crust is of considerable ecological importance in insulating the internal airspace system from the external climatic extremes. In the case of Antarctic deserts, this means that lichens require no unusual physiological adaptation though morphogenetic adaptation to a filamentous form is required to permit the lichen to penetrate the porous substrate. Lichens are found only in cold desert environments, these organisms, and indeed all eukaryotes, appear unable to cope with the environmental stresses associated with endolithic life in hot, extremely arid, deserts. Cryptoendolithic micro-organisms inhabit a 'closed ecosystem' in which a steady state exists where photosynthetic activity is balanced by respiration. It has been stated that cryptoendolithic communities are among the most highly closed ecosystems in nature (Friedmann and Ocampo-Friedmann, 1984).

Epilithic life is not restricted to the terrestrial environment, such micro-organisms colonizing rocks in both fresh and marine waters. Epilithic micro-organisms are found in humid conditions in caves and are also involved in the colonization of masonry. Epilithic micro-organisms are associated with the formation of 'desert varnish' on rocks.

Deep, subsurface communities of micro-organisms

For practical reasons there have been relatively few studies of micro-organisms in deep, subsurface habitats although it is well established that significant populations exist at depths in excess of 400 m. The size and nature of populations is dictated by the nature of the substrata but, in general, communities are diverse and metabolically active, consisting of aerobic, micro-aerophilic and anaerobic micro-organisms. These micro-organisms are identical, or closely related, to their counterparts at the terrestrial surface. Activities are greatest in water-saturated, subsurface sands and are least in relatively impermeable clay zones. Methophils are particularly numerous in some groundwaters where optimal conditions for growth occur due to the presence of methane diffusing from anaerobic sediments, where methanogenic activity can be considerable, and of oxygen diffusing from the atmosphere. Methophils may, however, be sensitive to common pollutants such as halogenated aliphatic hydrocarbons (e.g. dichloromethane). Metabolically active, sulphate-reducing bacteria have also been isolated from deep, subsurface sediments despite the low nutrient levels generally assumed to be present. Sulphate-reducing bacteria in sandy aquifer sediments in the London basin were able to utilize a variety of organic substrates while sulphate reduction was supported by an indigenous organic source within the sediments. In this situation, however, sulphate reduction will be limited by the presence of O_2, the quantity and quality of organic carbon available and, possibly, SO_2. Sulphate-reducing bacteria may play a beneficial role in the degradation of recalcitrant organic compounds. There is a limited effect on groundwater quality through small increases in iron concentration (Johnson and Wood, 1993).

The limitation of bacterial populations occurs by protozoal grazing, as in surface soils, and it is presumed that bacteriophages are also present. Anaerobic protozoa are present in anoxic sediments, usually being members of the Archezoa. Anaerobic, free-living, ciliated protozoa (275–277), including species of *Caenomorpha*, *Metopus* and

275–277 Free-living, anaerobic, ciliated protozoa and symbiotic methanogens. A living cell of *Metopus palaeformis* (**275**, x 290) alongside the same ciliate (**276**, x 390) showing the presence of autofluorescing methanogens. A confocal image of a dual-probed (bacteria/Archeae) ciliate, *Cyclidium porcatum* (**277**, x 570), illustrates a new type of symbiosis. Bacteria fluoresce green and Archeae (methanogens) red within the intracellular tripartite complex. The red-fluorescing clusters surrounding the ciliate are *Methanosarcina*. Reproduced with permission from Embley, T.M. and Finlay, B., 1994, © 1994, The Society for General Microbiology.

Sanderia, have been isolated from anaerobic, sulphide-rich sands (Fenchel *et al.*, 1977). These protozoa contain methanogenic bacteria as endosymbionts. Mitochondria and cytochrome oxidase are lacking but hydrogenosomes, specialized redox organelles present in other anaerobic protozoa and fungi, are present. Hydrogenosomes oxidize pyruvate with the production of ATP, CO_2, H_2 and acetate. This H_2-evolving fermentation appears to be the basis of the symbiosis, methanogens being shielded from competition with sulphate-reducing bacteria (*Feature 13*).

Feature 13. Deep six

Deep, subsurface micro-organisms are of considerable potential importance in the bioremediation of terrestrial sediments and aquifers polluted by organic and radioactive wastes. The populations, as a whole, have considerable metabolic versatility and are able to degrade a wide range of organic compounds including hydrocarbons, aromatic compounds and chlorinated aliphatic compounds. The latter category includes compounds, such as trichlorethylene, that are highly recalcitrant and serious environmental pollutants. Methanogens can metabolize a range of pollutants and are of importance in this role in a number of environments. Studies of consortia degrading toluene (**278**) and *o*-xylene (**279**) showed a high degree of substrate specificity and a long adaptation period. Pollutants may not be degraded in the presence of more readily metabolized co-contaminants. DNA repair mechanisms are highly conserved in bacteria in deep, subsurface communities and appear to be of importance in maintaining the integrity of DNA and providing protection against chemical insults, such as oxygen radicals, during periods of slow growth. From Arrage *et al.*, 1993.

278, **279** A toluene-degrading methanogenic consortium (**278**) was composed primarily of two distinct rod-shaped bacteria, one with rounded ends, the other with blunt ends abutting on to the next cell in the chain. These latter bacteria appeared similar to *Methanothrix soehngenii*. An *o*-xylene-degrading consortium (**279**) had similar constituents although the round-ended rods were fatter and more ellipsoidal. A thin rod was also more prevalent and a notable feature of the consortium was the web of exopolysaccharide-like material. Reproduced with permission from Edwards, E.A. and Grbic-Galic, D., 1994. © 1994 American Society for Microbiology.

Metal ores

Some micro-organisms, including bacteria, algae and lichens, are able to live in ores containing high metal concentrations. In some cases, the organism is highly tolerant of high concentrations of metals but has no other interaction with the metal. This situation tends to favour microfungi and the actino-mycetes. In many cases the metal is solubilized as a consequence of microbial metabolism. Solubilization may be beneficial both to the active micro-organism and to other organisms due to the mobilization of nutritionally important metals. In some cases immobilization leads to an increase in concentration of metals to a toxic level.

An unusual relationship occurs between penicillin-producing fungi and *Bacillus cereus* in areas containing high metal concentrations. Penicillin-producing fungi are highly resistant as a consequence of the breakdown of penicillin into the metal-complexing substance penicillamine. *Bacillus cereus* is not inherently resistant to metals but is highly penicillin resistant and is able to use the fungal defence mechanism against metal toxicity.

Solubilization may be direct (due to enzyme-mediated activity) or indirect (due to the production of corrosive metabolites such as acids, ligands and NH_3). Direct attack involves either oxidation or reduction of the metal and is usually beneficial to the micro-organism. These can only be prokaryotes since the electron transport system is located in the cell envelope and not in the intracellular organelles. Oxidative reactions provide a partial or total source of energy and are themselves of two types, direct or indirect. Oxidative reactions are beneficial in that the mineral serves as either a partial or complete energy source. In reducing processes, the mineral serves, to a greater or lesser extent, as the electron sink for respiration.

Direct attack on ores is of considerable economic importance in that the process can be exploited as a means of extracting minerals from low-grade mined materials and also in mining itself.

Indirect attack on metal ores occurs when meta-bolic products react with an ore and cause eventual solubilization. This involves several possible mechanisms (Ehrlich, 1992).

A number of micro-organisms also have sequestering properties for metals in solution. These are important in an overall sense in that sequestration may be the nucleation process for *de novo* mineral deposition in nature. Sequestration is also important as a means of removing toxic metal waste, including radioactive metals, from solutions. The process is also involved in the flotation method for the recovery of metals from ores. Micro-organisms may also be used as a means of locating ore deposits. The basis is to search for micro-organisms known to be associated with ore deposits. A number of approaches are possible (*Table 6*).

Table 6. A miner forty-niner – use of micro-organisms in prospecting for metal ores.

Enrichment methods for detection of index organisms
Use of DNA probes for detection of index organisms
Detection of bacteria with unusually high resistance to ores sought
Survey for high concentration of *B. cereus*

REFERENCES

Ambus, P. (1993) Control of denitrification enzyme activity in a streamside soil, *FEMS Microbiology Ecology*, **10**, 225–34.

Arrage, A.A., Phelps, T.J., Benoit, R.E. and White, D.C. (1993) Survival of subsurface microorganisms exposed to ultraviolet irradiation and hydrogen peroxide, *Applied and Environmental Microbiology*, **59**, 3345–50.

Astrom, B., Gustafsson, A. and Gerhardson, B. (1993) Characteristics of a plant deleterious rhizosphere pseudomonad and its inhibitory metabolite(s), *Journal of Applied Bacteriology*, **74**, 20–8.

Blakeman, J.P. (1991) Foliar bacterial pathogens: epiphytic growth and interactions on leaves, *Journal of Applied Bacteriology (Supplement)*, **70**, 49–59.

Blakeman, J.P. and Brodie, I.D.S. (1977) Competition for nutrients between epiphytic micro-organisms and germination of spores of plant pathogens on beetroot leaves, *Physiological Plant Pathology*, **10**, 29–42.

Bock, E. and Sand, W. (1993) The microbiology of masonry deterioration, *Journal of Applied Bacteriology*, **74**, 503–14.

Carter, J.P., Hsiao, Y.H., Spiro, S. and Richardson, D.J. (1995) Soil and sediment bacteria capable of aerobic nitrogen respiration, *Applied and Environmental Bacteriology*, **61**, 2852–8.

Collins, C.H. and Kennedy, D.A. (1992) The microbiological hazards of municipal and clinical wastes, *Journal of Applied Bacteriology*, **73**, 1–6.

Edwards, E.A. and Grbic-Galic, D. (1994) Anaerobic digestion of toluene and o-xylene by a methanogenic consortium, *Applied and Environmental Microbiology*, **60**, 313–22.

Ehrlich, H.L. (1990) *Geomicrobiology*, Marcel Dekker, New York.

Ehrlich, H.L. (1992) Metal extraction and ore discovery, *Encyclopedia of Microbiology* (ed. J. Lederberg), pp. 75–80, Academic Press, New York.

Embley, T.M. and Finlay, B. (1994) The use of small subunit rRNA sequences to unravel the relationship between anaerobic ciliates and their methanogenic endosymbionts, *Microbiology*, **140**, 225–35.

Evans, G.M. (1993) The use of microorganisms in plant breeding, *Exploitation of Microorganisms* (ed. D.G. Jones), pp. 225–48, Chapman & Hall, London.

Fenchel, T., Perry, T. and Thane, A. (1977) Anaerobiosis and symbiosis with bacteria in free-living ciliates, *Journal of Protozoology*, **24**, 154–63.

Fetzer, S., Bak, F. and Conrad, R. (1993) Protective role of pyrites (FeS_2) and survival of methanogenic bacteria, *FEMS Microbiology Ecology*, 12, 107–115.

Fiddaman, P.J. and Rossall, S. (1993) The production of antifungal volatiles by *Bacillus subtilis*, *Journal of Applied Bacteriology*, **74**, 119–26.

Friedmann, E.I. and Ocampo-Friedmann, R. (1984) Endolithic microorganisms in extreme dry environments: analysis of a lithobiontic microbial habitat, *Current Perspectives in Microbial Ecology* (ed. C.A. Reddy), pp. 177–85, American Society for Microbiology, Washington DC.

Galloway, D.J. (1992) Biodiversity: a lichenological perspective, *Biodiversity and Conservation*, **1**, 312–23.

Gianinazzi-Pearson, V. and Gianinazzi, S. (1989) Phosphorus metabolism in mycorrhizas, *Nitrogen, Phosphorus and Sulphur Utilisation by Fungi* (eds. L. Boddy, R. Marchant and D.J. Read), pp. 227–43, CUP, Cambridge.

Hawksworth, D.L. (1988) Coevolution of fungi with algae and cyanobacteria in lichen symbioses, *Coevolution of Fungi with Plants and Animals* (eds. K.A. Pirozynski and D.L. Hawksworth), pp. 125–48, Academic Press, London.

Hawksworth, D.L. (1990) The long-term effects of air pollutants on lichen communities in Europe and North America, *The Earth in Transition: Patterns and Processes of Biotic Impoverishment* (ed. G.M. Woodwell), pp. 45–64, CUP, Cambridge.

Hawksworth, D.L., Coppins, B.J. and Rose, F. (1974) Changes in the British lichen flora, *The Changing Flora and Fauna of Britain* (ed. D.L. Hawksworth), pp. 47–78, Academic Press, London.

Hirano, S.S. and Upper, C.D. (1983) Ecology and epidemiology of foliar bacterial plant pathogens, *Annual Review of Phytopathology*, **21**, 243–69.

Hirano, S.S. and Upper, C.D. (1988) Rain and the population dynamics of *Pseudomonas syringae* on snap bean leaflets, *Proceedings of the 5th International Congress of Plant Pathology*, p. 86, Kyoto.

Johnson, A.C. and Wood, M. (1993) Sulphate-reducing bacteria in deep aquifer sediments of the London basin: their role in anaerobic mineralization of organic matter, *Journal of Applied Bacteriology*, **75**, 190–7.

Jordan, D.C. (1984) *Rhizobium, Bergey's Manual of Systematic Bacteriology*, Vol. 1 (eds. N.R. Krieg and J.G. Holt), pp. 235–42, Williams and Wilkins, Baltimore.

Kusel, K. and Drake, H.L. (1995) Effects of environmental parameters on the formation and turnover of acetate by forest soils, *Applied and Environmental Microbiology*, **61**, 3667–75.

Lewis, D.H. (1987) Evolutionary aspects of mutualistic associations between fungi and photosynthetic organisms, *Evolutionary Biology of the Fungi* (eds. A.D.M. Rayner, C.M. Brasier and D. Moore), pp. 161–78, CUP, Cambridge.

Marsh, P., Toth, I.K., Meijer, M. *et al.* (1993) Survival of the temperate actinophage OC31 of *Streptomyces lividans* in soil and the effects of competition and selection on lysogeny, *FEMS Microbiology Ecology*, **13**, 13–22.

McCormack, P.J., Wildman, H.G. and Jeffries, P. (1994) Production of antibacterial compounds by phyllophane-inhabiting yeasts and yeastlike fungi, *Applied and Environmental Microbiology*, **60**, 927–31.

McCurdy, H.D. (1986) Myxococcales, *Bergey's Manual of Systematic Bacteriology*, Vol. 3 (eds. P.H.A. Sneath, N.S. Mair, M.E. Sharpe and J.G. Holt), pp. 2139–2148, Williams and Wilkins, Baltimore.

Mitchell, D.T. (1993) Mycorrhizal associations, *Exploitation of Microorganisms* (ed. D.G. Jones), pp. 169–96, Chapman & Hall, London.

Mormile, M.R., Gurijala, K.R., Robinson, J.A. *et al.* (1996) The importance of hydrogen in landfill fermentations, *Applied and Environmental Microbiology*, **62**, 1583–8.

Moser, E., Goretzki, L. and Bock, E. (1990) Hohe stickstoffdeposition an gebaudeoberflachen in Wittenberg, *Bautenschutz Bausanierung*, **13**, 98–101.

Owens, J.D. and Keddie, R.M. (1969) The nitrogen nutrition of soil and herbage coryneform bacteria, *Journal of Applied Bacteriology*, **32**, 338–45.

Pacovsky, R.S. (1990) Development and growth effects in the *Sorghum–Azospirillum* association, *Journal of Applied Bacteriology*, **68**, 555–63.

Poelt, J. (1991) Homologies and analogies in the evolution of lichens, *Frontiers in Mycology* (ed. D.L. Hawksworth), pp. 85–97, CAB International, Wallingford.

Read, D.J. (1991) Mycorrhizas in ecosystems – nature's response to the 'Law of the Minimum', *Frontiers in Mycology* (ed. D.L. Hawksworth) pp. 101–30, CAB International, Wallingford.

Richardson, D.H.L. (1991) Lichens and man, *Frontiers in Mycology* (ed. D.L. Hawksworth), pp. 187–210, CAB International, Wallingford.

Roberson, E.B. and Firestone, M.K. (1992) Relationship between desiccation and exopolysaccharide production in a soil *Pseudomonas* sp., *Applied and Environmental Microbiology*, **58**, 1284–91.

Sarkoh, N.A., Al-Hasan, R.H., Khaferer, M. and Radwan, S.S. (1995) Establishment of oil-degrading bacteria associated with cyanobacteria in oil-polluted soil, *Journal of Applied Bacteriology*, **78**, 194–9.

Schreiner, D.R. and Koide, R.T. (1993) Stimulation of a vesicular–arbuscular mycorrhizal fungi by mycotrophic and nonmycotrophic plant root systems, *Applied and Environmental Microbiology*, **59**, 2750–2.

Schuler, R. and Haselwandter, K. (1988) Hydroxamate siderophore production by ericoid mycorrhizal fungi, *Journal of Plant Nutrition*, **11**, 907–13.

Smith, R.I.L. (1990) Signy Island as a paradigm of biological and environmental change in antarctic terrestrial ecosystems, *Antarctic Ecosystems – Ecological Change and Conservation* (eds. K.R. Kerry and G. Hempel), pp. 32–50, Springer-Verlag, Berlin.

Truper H.G. and Galinski, E.A. (1986) Concentrated brines as habitats for micro-organisms, *Experientia*, **42**, 1182–7.

Tsuji, T., Kawasaki, Y., Takeshima, S. *et al.* (1995) A new fluorescence technique for visualizing living microorganisms in soil, *Applied and Environmental Microbiology*, **61**, 3415–21.

Vazquez, F.J., Acea, M.J. and Carballas, T. (1993) Soil microbial populations after wildfire, *FEMS Microbiology Ecology*, **13**, 93–104.

Wachtershauler, G. (1992) Groundworks for an evolutionary biochemistry: the iron–sulphur world, *Progress in Biophysics and Molecular Biology*, **58**, 85–201.

Xu, L.-H., Li, Q.-R. and Cheng, L.-J. (1996) Diversity of soil actinomycetes in Yunnan, China, *Applied and Environmental Microbiology*, **62**, 244–8.

Yang, C.-H., Menge, J.A. and Cooksey, D.A. (1994) Mutations affecting hyphal colonisation and pyoverdine production in pseudomonads antagonistic towards *Phytophthora parasitica*, *Applied and Environmental Microbiology*, **60**, 473–81.

4 EXTREME ENVIRONMENTS

In the context of higher animal life many of the environments favoured by micro-organisms appear remarkably inhospitable. An environment, however, that permits the development of a wide diversity of micro-organisms, or that places only moderate selective pressure on its inhabitants is not regarded as extreme. Very low temperature environments, for example, are a common feature of life on this planet and are colonized by a wide range of organisms. In this chapter, four types of environment are considered extreme: very high temperature, highly acidic, highly alkaline and highly saline. There is considerable overlap in that many high temperature environments are also highly acidic and many highly alkaline environments are also highly saline. In each case conditions are unfavourable to the great majority of micro-organisms and a highly adapted and characteristic microflora develops.

Many of the micro-organisms developing under the most extreme conditions are archaebacteria (archaea). This has been attributed to conditions in extreme environments generally being similar to those that prevailed on earth in the earliest (archaeic) times.

Feature 14. An industrial revolution

There is considerable interest in the use of thermophilic micro-organisms in biotechnology. Fermentations proceed more rapidly at high temperatures and there is less concern over contamination. There has been particular interest in the use of thermophilic anaerobic fermentations in the production of fuel and chemicals (*Table 8*). Enzymes derived from thermophilic bacteria are also of interest due to their high level of heat stability. Considerable use is already made of thermostable enzymes, including a xylose isomerase from *B. coagulans* used in the manufacture of high-fructose corn syrup at 60°C (140°F). On a smaller scale, thermostable DNA polymerase is produced for use in the polymerase chain reaction; the use of heat-stable proteins from the same source has been proposed in the manufacture of semiconductor chips.

Microbial communities in thermophilic environments

Hydrothermal vents and seamounts

The hydrology of hydrothermal vents has been discussed in the context of the ocean as a whole in 'Deeps-sea habitats' (see pages 40–42). Thermophiles are important as both producers and consumers of organic matter. Among the more important genera are *Pyrodictium occultum*, *Pyrococcus* and *Thermococcus* as well as poorly defined archaebacterial isolates. *Archaeoglobus* appeared to be present only in a hydrothermal system off Italy, but has now also been isolated from a volcanic slick from an erupting Polynesian seamount and from water in a geothermally heated North Sea oil well (see below). An open-sea volcanic plume from the erupting MacDonald seamount contained at least 10^6 hyperthermophiles per litre, a number of which were closely related to existing archaebacteria including

Archaeoglobus (Haber *et al.*, 1990). This was considered to provide evidence of spreading and long-term survival of hyperthermophiles at temperatures below those supporting growth. Some isolates were, however, unique to the erupting seamount, including isolate MD7 (**280**), an obligate anaerobe recovered by incubation at 100°C (212°F) for 24 hours under a pressure of 300 kPa.

Although the seamount ecosystem was not understood, it was thought that the production of organic matter resulted from the activity of the anaerobic H_2-oxidizers *Pyrodictium* and *Archaeoglobus*. These micro-organisms, which use CO_2 as the sole carbon source, obtain energy by sulphur and SO_4^{2-} respiration respectively. The seamount community also contains strictly heterotrophic anaerobes including *Pyrococcus*, *Thermococcus* and MD7. These organisms are consumers of organic matter, growing by fermentation, or sulphur

Table 8. Possible uses of thermophilic fermentations in biotechnology.

Production of methane from waste biomass by
 a methanogenic consortium
Production of ethanol by *Cl. thermocellum*
Synthesis of amino acids by *B. coagulans*
Production of antibiotics by *Thermoactinomyces*
Production of lactic acid by *Clostridium*,
 Thermoanaerobacter or *B. coagulans*
Production of acetic acid by *Clostridium* and
 Acetogenium
Synthesis of carotenoids by *Th. aquaticus*

280 Diagrammatic representation of hyperthermophilic isolate MD7 from an erupting seamount.

respiration, on proteins, sugars and the like. (Haber *et al.*, 1990). Synthesis of amylolytic enzymes in response to complex carbohydrates has been reported in *Pyrococcus*; polysaccharide availability appears to be an important aspect of growth in hydrothermal environments.

Hot springs

Hot springs and nearby soils vary both in temperature and pH value, with obvious consequences for the microbial communities developing. Light intensity is also of importance where photosynthetic microorganisms are present. In non-acid springs, at temperatures below 74°C (165°F), cyanobacteria are of major importance in primary production and, in the absence of grazing by predators, microbial mats develop. Like microbial mats in other environments, these exhibit both vertical and temporal heterogeneity. Cyanobacteria and green non-sulphur bacteria dominate the upper 1 mm but are not usually present at depths below 2.5 mm, probably as a result of phototrophic behaviour. *Synechococcus* is the most common culturable cyanobacterium while the dominant culturable green non-sulphur bacterium is the filamentous gliding bacterium *Chloroflexus*. *Synechococcus* and *Chloroflexus* grow in close association, although *Synechococcus* is favoured by mildly acid conditions and *Chloroflexus* by mildly alkaline conditions and higher temperatures. During growth in the light, *Synechococcus* accumulates carbohydrates, which are then used during growth in the dark. Some 10% of photosynthetically fixed carbon is excreted by *Synechococcus* as organic carbon, which is available to *Chloroflexus*. Some of the metabolic transformations which occur in the light may be due to *Chloroflexus* and a low level of heterotrophic uptake is possible in the dark. Irrespective of the mechanism, accumulation of C-storage molecules in the light can be used to drive the synthesis of protein in the dark. Photosynthetic activity by *Chloroflexus* is limited by the presence of oxygen in upper layers of the mat and consequent repression of bacteriochlorophyll synthesis.

Anaerobic, thermophilic, glycolytic bacteria have been isolated from various hot springs. The major nutrient source appears to be xylans, derived from vegetation blown into the spring. Most of the xylan-utilizing bacteria are Gram-negative, asporogenous, non-motile rods, identified with *Dictyoglomus*.

The situation in hot, acid springs is different, in that cyanobacteria and green non-sulphur bacteria

281 A diagrammatic representation of *Sulfolobus*.

are unable to grow and mats do not develop. Environments of this type are dominated by hyperthermophiles, especially *Sulfolobus* (**281**). This organism has an optimum pH value of *c*.2.0 although growth occurs over the range from pH 1.0 to 5.9. In hot springs and their immediate surroundings, *Sulfolobus* grows as a respiratory chemoautotroph. The principle substrate for growth is elemental sulphur, the bacterium growing on the surface of crystalline sulphur and oxidizing it to sulphuric acid. Elemental sulphur is formed in hot springs by the leaching of iron and sulphide by geothermal steam or hot water. Sulphide is then oxidized to elemental sulphur by either chemical or biological processes involving oxygen or ferric iron. *Sulfolobus* is very widely distributed in hot springs, but other thermoacidophiles appear to have a limited distribution. Anaerobic archaebacteria of the *Thermoproteus* group, for example, have been isolated only from geothermal regions of Iceland. This group originally comprised two genera (the rod-shaped *Thermoproteus* and the coccal *Desulfurococcus*) and has a minimum pH value for growth of 5.0 and is not, therefore, found in strongly acid springs. In nature, the organisms, despite being strict anaerobes, have a predominantly respiratory metabolism with elemental sulphur serving as the terminal electron acceptor. More recently, a related bacterium, *Pyrobaculum aerophilum*, has been described which is capable of aerobic growth or nitrate respiration.

Geothermally heated oil wells

These have much in common with hydrothermal systems in terms of their physics and chemistry. Wells are typically drilled to a depth of 2–4 km where temperatures are in the range 60–130°C (140–266°F) and pressures between 20–50 mPa. Most wells contain formation water (*in situ* pore water) that has a low SO_4^{2-} content and does not support significant growth of sulphate-reducing bacteria. It is common practice, however, to enhance oil recovery by flooding wells with anoxic seawater, which may stimulate thermophilic, sulphate-reducing bacteria capable of growth on organic acids. The resultant H_2S production can cause severe problems, through corrosion of metal alloys used in wells and processing equipment. Precipitation of sulphides may cause extraction problems while H_2S itself reduces oil quality by souring and presents a health hazard to personnel working on production platforms.

Thermophilic strains of *Desulfotomaculum* are widespread and *Thermodesulfurobacterium* are also common. A new species of *Archaeoglobus*, *Arch. fulgidus*, has been isolated from high temperature wells where it probably grows in the well-head and oil–water separator at temperatures between 70 and 85°C (158 and 185°F). In moderately thermophilic wells (65–75°C (149–167°F)) all three species are able to coexist, possibly growing in association with fermentative bacteria producing H_2 and valerate for mixotrophic growth. Although the sulphate content is low, sulphate accumulation and high-affinity uptake permits growth.

Self-heating coal refuse

Coal refuse containing significant quantities of residual coal and iron pyrites (FeS) is prone to heating due to the oxidation of FeS by chemoautotrophic bacteria. This also results in acidification, creating suitable conditions for the growth of the enigmatic bacterium *Thermoplasma*. This organism is found only in coal refuse with a temperature between 37 and 65°C (98.6 and 149°F) and a pH value between 0.5 and 4.5. *Thermoplasma* has not been isolated from any naturally occurring hot environment and it has been speculated that the natural habitat is coal veins. Coal, pyrolysed by heating, supplies *Thermoplasma* with growth factors. The nature of these is not known but the requirement can be satisfied by an oligopeptide from yeast extract.

Charcoal piles

Charcoal piles provide an eco-niche to a highly specialized and adapted thermophilic actinomycete, *Streptomyces thermoautotrophicus*, a CO$^-$ and H$_2^-$ oxidizing bacterium. Charcoal piles consist of burning wood covered with a layer of soil. The temperature profile of the soil closely matches that of the organisms' growth while smoke provides a constant source of CO and H_2. Condensing water vapour maintains the soil in a moist condition and minerals are obtained from ashes. Charcoal burning is an intermittent process but *Str. thermoautotrophicus* has at least two mechanisms permitting extended survival. Exospores are produced, which withstand desiccation and temperatures up to 100°C (212°F). At the same time, oxidation of CO at temperatures down to 10°C (50°F) allows the organism to meet its maintenance requirements although not to grow (Gadkari *et al.*, 1990).

EXTREMELY ACIDIC ENVIRONMENTS

Occurrence and distribution of extremely acidic environments

Although many environments, including some foods, have pH values significantly below neutrality, these are not regarded as extremely acidic. Extremely acidic environments (pH value $c.3.0$ or less) are not widespread in nature – the most common are strip coal mining operations, coal waste heaps, metal ore spoil heaps and drainage water from ore mines. These environments are typically high in sulphate, pyrite and H^+ ions. Acidity is derived from the oxidation of sulphides and pyritic material to sulphuric acid. Above pH 4.5 abiotic oxidation occurs at a slow rate. Below this pH oxidation is mediated by bacteria and proceeds at a rate some 10^6 times faster. Some acid peat has a pH value of $c.2.0$ and is also considered extremely acidic. As discussed above (see page 133) some hot springs and surrounding acid soils are also extremely acidic and populated by thermoacidophiles. Extremely acidic environments are found in association with chemical processing including laboratory reagents. In general, however, no attempt has been made to study the micro-organisms present.

Acidophilic and acid-tolerant micro-organisms

Micro-organisms able to grow in extremely acidic environments comprise both acidophiles and acid-tolerant types. Various definitions of acidophile exist, but in the current context these are considered to be micro-organisms that grow optimally at a pH of 3.0 or less but not close to neutrality. The majority of species of acidophiles are thermo-acidophilic archaebacteria and so are discussed in the context of high temperature environments. Some of these micro-organisms, however, are of very restricted distribution. Non-thermophilic acidophiles are eubacteria which, although belonging to a smaller number of genera, are much more widespread and more versatile. The best known genus is *Thiobacillus* which has a pH_{opt} of $c.2.0$. The two most common species are *Th. thiooxidans* (which utilizes elemental sulphur as its energy source) and *Th. ferrooxidans* (which can utilize both reduced sulphur compounds and Fe^{2+}). *Th. ferrooxidans* is of considerable interest in the recovery of metals from ores by bioleaching (see below). Moderate thermophiles resembling *Th. ferrooxidans* have also been isolated, while a second Fe^{2+}-oxidizing thiobacillus, *Th. prosperus*, is a moderate halophile. In the past *Thiobacillus* and other members of the genus have often been considered to be obligate chemolithotrophs but a mixotrophic strain of *Th. ferrooxidans* has been described and *Th. acidophilus* and *Th. cubrinus* both have a mixotrophic metabolism, obtaining energy by the oxidation of reduced sulphur compounds but not Fe^{2+}.

Leptospirillum ferrooxidans is a second species of iron-oxidizing acidophile which, in contrast to *Th. ferrooxidans*, is incapable of oxidizing reduced sulphur compounds. Some strains are moderately thermophilic. Moderately thermophilic, iron-oxidizing, endospore-forming bacteria have also been isolated. In contrast to *Leptospirillum* and *Thiobacillus*, these bacteria can grow as chemolithoheterotrophs, obtaining energy from Fe^{2+} oxidation and carbon from organic substrates.

Heterotrophic acidophiles were first isolated fairly recently. The best characterized genus is *Acidophilium*; this grows in close association with *Thiobacillus* and is a common contaminant in cultures of thiobacilli. *Acidophilium* shares a number of properties with *Thiobacillus*, although there are significant differences in carbon and energy metabolism. Heterotrophic acidophiles have also been isolated which resemble neutrophilic bacteria of the *Sphaerotilus–Leptothrix* group in morphological and behavioural characteristics but which grow in the pH range 2.0–4.4.

Although most eukaryotic micro-organisms in low pH environments are acid-tolerant rather than acidophilic, the algae *Chlamydomonas acidophila* and *Euglena mutabilis* reportedly each have a pH_{opt} of $c.3.0$ and a pH_{max} of $c.5.5$.

Acid-tolerant organisms are those able to grow

at low pH values but which have a significantly higher pH_{opt}. Relatively few non-acidophilic bacteria are capable of good growth at pH values below 4.0 and, with important exceptions such as *Metallogonium*, are not found in the same habitats as acidophiles. Yeasts are able to grow at pH values as low as 2.0 while fungi have the widest pH growth range, strains of *Aspergillus*, *Fusarium* and

Penicillium reportedly growing over the range pH 2–10. Acid-tolerant algae are known, some species of *Chlorella* are able to grow at pH 2.0 but have a pH_{opt} of $c.7.0$. Flagellate protozoa, resembling *Eutrepia* and several amoeboid protozoa have been isolated from water at pH 2.0 and while actual growth was not demonstrated for a number of years, grazing has now been observed.

Microbial communities in extremely acidic environments

Ore spoil heaps and mine drainage water

Development of acidity in ore spoil heaps initially involves the slow abiotic formation of sulphuric acid followed by a microbial succession in which *Metallogonium* lowers the pH value sufficiently for growth of *Thiobacillus*.

Although *Thiobacillus* is numerically dominant it is found in close association with *Acidophilium* and, where sufficient Fe^{2+} is available, *Leptospirillum*. There is evidence of adaptation to the ambient temperatures at the mine site, in that psychrotrophic strains of *Thiobacillus* and *Acidophilium* have been isolated from a Canadian uranium mine where temperatures do not usually exceed 18°C (64.4°F). At the same time moderately thermophilic strains of *Thiobacillus* and *Leptospirillum* have been isolated from heated spoil heaps. Such isolates, however, are able to grow over a wide temperature range, including higher ambient temperatures. In addition to acido-

philic bacteria, yeasts resembling *Rhodo-torula* and various protozoa have been observed in acidic mine drainage water although numbers are generally small.

Heterotrophic acidophiles, apparently related to the *Sphaerotilus–Leptothrix* group, were isolated from a disused pyrites mine in Wales. The mine was characterized by the copious growth of gelatinous 'acid streamers' in the main drainage stream and on moist surfaces. These consisted of a mixed community of filamentous and unicellular bacteria, the latter probably being *Th. ferrooxidans*, *Ls. ferrooxidans* and *Acidophilium*. The bacteria were being grazed by protozoa and rotifera (*Feature 15*).

Micro-organisms associated with ores such as copper require a high tolerance of the metal ions. Metal resistance does not appear to be an inherent property of the micro-organisms, but results from reduced interaction with the metals. The mechanism involves competition for metal-sensitive sites between toxic metal cations and the excess H^+ ions usually present (Brierley *et al.*, 1980).

Feature 15. The banks of the Ohio

The large quantities of sulphuric acid produced by *Thiobacillus* inevitably enter drainage water and are a serious source of pollution. During the 1960s it was estimated that 3×10^4 tons of sulphuric acid entered the Ohio River annually as a result of pollution from mine drainage. This is probably an extreme example but localized acid pollution undoubtedly occurs elsewhere. In Wales, for example, acid pollution by water emerging from the addit of a long-disused pyrites mine can be seen to affect the immediate environment, the stream containing massive growths of a *Gallionella*-like organism. *Gallionella* is not, however, usually associated with acidic environments (see also above).

Ore bioleaching operations

Bioleaching involves the use of micro-organisms to solubilize the metals contained in mineral ores. This enables the extraction of the mineral as an aqueous solution. Bioleaching is environmentally attractive since there is no air pollution, although engineering precautions must be taken to prevent contamination of groundwater. The process makes possible the recovery of metal from low-grade (lean) ores, metal that cannot be recovered by conventional pyrometallurgical processes because of an unfavourable ratio of host rock to ore mineral. The process is also economically attractive because it is neither energy nor labour intensive. Bioleaching is currently used for the solubilization of sulphidic copper and uranium ores, although extension to other ore types and other metals is perfectly feasible.

Thiobacillus ferrooxidans is the most important organism in bioleaching although *Th. thiooxidans* and *Ls. ferrooxidans* are also involved. *Acidophilium* is present in leaching operations and is thought to enhance activity of *Th. ferrooxidans* by removing inhibitory organic compounds. *Acidophilium* may also play a direct role in leaching through its apparent ability to oxidize elemental sulphur in the presence of an organic energy source. *Acidophilium* is also able to reduce Fe^{3+} to Fe^{2+} and thus regenerate an energy source for *Th. ferrooxidans* while contributing to the leaching process. Other micro-organisms, including fungi and protozoa, are present in the community which develops but relationships and consequences for metal recovery are little understood.

Ore leaching processes are of two types: *in situ* leaching of the ore in place and ore leaching after recovery by conventional mining. In each case a 'barren' solution containing ferric sulphate and sulphuric acid is allowed to percolate through the ore body. This creates conditions which enrich for an acidophilic microflora. In some cases, however, the process is accelerated either by recycling barren fluid containing acidophilic micro-organisms or by inoculating the barren fluid with desired micro-organisms. Conditions in ore leaching environments are highly favourable for growth of acidophiles although a particularly high level of metal tolerance is required. Under some circumstances there is a considerable degree of heating, favouring the growth of moderately thermophilic strains and, in localized areas, temperatures may be sufficiently high to support growth of SO- and Fe^{2+}-oxidizing strains of *Sulfolobus*. A further use of acidophiles in metallurgy is the biobeneficiation of gold ores. This involves use of *Th. ferrooxidans* to remove pyrites and arsenopyrites from sulphidic gold ores. This prevents interference from the pyrites during chemical extraction of the gold.

Strip coal mining and coal waste heaps

As an environment for acidophiles, pyrites-containing coal and coal waste resembles ore spoil heaps and the microflora is generally similar. Acid-tolerant fungi are more prevalent and can be isolated both from coal waste and drainage water at pH 2–3. A considerable degree of heating may occur, creating suitable conditions for *Thermoplasma* (*Feature 16*).

Feature 16. Down the pit

Acidophilic bacteria cause problems in strip mining operations, causing a loss of coal quality. Conversely, acidophiles have been proposed as a means of desulphurizing coal and thus reducing SO_2 emissions and acid rain. At present, however, the large-scale exploitation of acidophiles in this way does not seem practical due to the recalcitrant nature of organic sulphur compounds. It is also argued that desulphurization cannot be justified economically and that large scale electricity generation, a major source of SO_2, should be progressively switched to alternative energy sources. These include nuclear power, use of other fossil fuels (such as natural gas), wind and tidal power.

HIGHLY ALKALINE ENVIRONMENTS

Occurrence and distribution of highly alkaline environments

Stable alkaline environments (pH > 10) arise where a combination of ecological, geographical and climatic conditions occur, which mitigate against the buffering effect of atmospheric CO_2. Soda lakes and deserts are found in many parts of the world including North, Central and South America, Australia and in many parts of Africa. In Asia soda lakes occur in Turkey, India and China and in Europe in Hungary, Yugoslavia and the former USSR. Springs, of high Ca^{2+} content are a further type of naturally occurring, highly alkaline environment found in a number of geographical locations, including California, Oman, Cyprus, the former Yugoslavia and Jordan.

A number of unstable, man-made, alkaline environments are associated with use of alkalis (usually NaOH) as part of an industrial process or are derived from the generation of $Ca(OH)_2$. Alkali is used in a wide range of industrial processes including electroplating, food manufacture and paper making, while $Ca(OH)_2$ is generated during cement manufacture and in blast furnace waste. Also, transient, highly localized, alkaline niches arise from biological activity in many types of environment, such as soils, which are classified on the basis of overall pH as neutral or even acidic.

Alkaliphilic and alkalitolerant micro-organisms

A number of definitions may be applied to micro-organisms from highly alkaline environments. Alkaliphiles may be defined as having a pH_{opt} for growth of more than 8.0. A more common pH_{opt} is from 9.0–10.0 while some species grow at pH values as high as 11.5 (*Feature 17*). A number of micro-organisms are able to grow at relatively high pH values but have a pH_{opt} of 7.0 or below. These micro-organisms are referred to as alkali-trophic or alkalitolerant. Alkaliphiles that are unable to grow at pH 7.0 are classified as obligate alkaliphiles while a further subdivision is made into halophilic alkaliphiles (haloalkaliphiles), also able to grow at NaCl concentrations up to saturation, and nonhalophilic alkaliphiles. Haloalkaliphiles are archaebacteria and only two genera are currently recognized, *Natronobacter* and *Natronococcus*. Other micro-organisms capable of growth at high pH values, however, are of much greater diversity.

Feature 17. Washday blues

Alkaliphiles are of commercial interest in the production of enzymes, such as alkaline protease, which are stable at high pH values. Alkaline proteases are widely used in biological washing powders. Industrial use is also made of a cyclodextrin glucanotransferase in the manufacture of cyclodextrin. This enzyme has activity over the pH range 4.5–10 and a conversion rate in excess of 75–80%, compared with 50% for the *B. macerans* enzyme conventionally used. This permits crystallization of the dextrin directly from the primary starch–enzyme reaction mixture. It should be appreciated, however, that alkaline enzymes are present in some neutrophilic bacteria, *B. licheniformis*, for example, being a source of some alkaline proteases used in washing powders.

Microbial communities in highly alkaline environments

Soda lakes and deserts

Soda lakes and deserts have a pH in the range 10.5–11.0. At these pH values, Ca^{2+} and Mg^{2+} precipitate as carbonates and so only very low levels are present in soluble form. Sodium carbonate is usually present as $Na_2CO_3 \cdot 10H_2O$ or $Na_2CO_3 \cdot Na_2CO_3 \cdot 2H_2O$ and there is a high, although variable, NaCl content. The best characterized soda lakes are those of the East African Rift Valley where both limnological and microbiological assays have been carried out for a number of years. The genesis of alkalinity in the Rift Valley has been attributed to vulcanism and soda-rich larva flows. Alternatively, alkalinity has been associated with sulphate reduction in anaerobic basins. These situations may not, however, be typical and, in general, soda lakes appear to arise from a particular combination of geological and climatic features (Grant, 1992). Low levels of Ca^{2+} and Mg^{2+} in the surrounding topography appear to be particularly important, together with the existence of distinct, isolated drainage basins and high rates of evaporation. These conditions are met in the Rift Valley which has a high Na^+, low Ca^{2+} and Mg^{2+} geology. Evaporative concentration has been of importance in the formation of the large number of lakes on the valley floor. These have a pH of 11.0–11.5 and vary in salinity from 5% in the north to saturation in lakes such as Lake Natron in the south. Soda lakes have a very high buffering capacity which maintains a constant high pH in conditions of varying rainfall. In some soda lakes, Ca^{2+} groundwater, which may have travelled a considerable distance, results in localized calcite ($CaCO_3$) precipitation and formation of 'tufa' columns. Cyanobacteria are involved in formation of such structures through the biogenic precipitation of aragonite, another crystalline form of $CaCO_3$.

Soda lakes have been described as the most productive naturally occurring aquatic environments, primary production being in excess of an order of magnitude greater than that of non-alkaline streams and lakes (Grant, 1992). To some extent this is due to the location of soda lakes in areas of high ambient temperature (30–35°C (86–95°F)) and a high intensity of illumination rather than to the specific properties of these environments. A specific factor in the very high productivity, however, is the virtually unlimited supply of HCO_3^- for photosynthesis. Prokaryotes are almost entirely responsible for primary production in soda lakes and blooms of phototrophs are common. To a large extent the NaCl content determines phototroph growth and more dilute lakes, with salinities of 5–15% NaCl, often exhibit permanent or seasonal cyanobacterial populations of only one or two genera at high cell densities. The dominant cyanobacteria are often poorly characterized but include the filamentous *Cyanospira* and the unicellular *Chroococcus* and *Pleurocapsa*. In some cases, blooms consist almost entirely of filamentous cyanobacteria referred to either as *Spirulina* (*Feature 18*) or *Arthrospira*. These cyanobacteria are gas vacuolated and float as a scum on the surface of less saline lakes. In some lakes *Spirulina* supports enormous populations of flamingos, Lake Nakura (Rift Valley) having an

---

Feature 18. Our daily bread

Spirulina is the only cyanobacterium consumed as human food, having been eaten by the Aztecs of Mexico and consumed today in Chad. In the western world, *Spirulina* is considered a health food due to its favourable amino acid content and low level of nucleic acids. In the sporting world, *Spirulina* had a vogue as a legal means of enhancing performance. This raises the question of whether it should be more permissive to use *Spirulina* than, for example, a steroid to improve performance. *Spirulina* is also of interest in biotechnology, its very high photosynthetic yield resulting in an efficient means of producing biochemicals such as essential amino acids and water-soluble vitamins.

282 Flamingos on Lake Nakura: a massive population supported by *Spirulina*.

283 A pink surface crust formed by *Natronobacterium* on the surface of Lake Magadi. **282** and **283** reproduced by courtesy of Prof. W.D. Grant, Department of Microbiology and Immunology, University of Leicester, UK.

estimated population of 2×10^6 birds consuming 200 tons of cyanobacterial biomass each day (**282**).

Eubacteria also play a role in primary production and the contribution of the anoxygenic phototroph *Ectothiorhodospira* is significant during this period; this bacterium blooms alongside cyanobacteria. The bacterium is also involved in the sulphur cycle through utilization of H_2S as electron donor in photosynthesis.

Less saline lakes also contain non-phototrophic bacteria at population densities of 10^7–10^8 ml^{-1}. Populations are stable and unaffected by cyanobacterial blooms or dilution after rainfall. The population is largely Gram-negative, aerobic bacteria, which cannot be assigned to existing taxa. Anaerobic non-phototrophs are also present and the presence of black anoxic sediments provides circumstantial evidence for the presence of alkaliphilic, sulphate-reducing bacteria. Methanogenesis is also a feature and is attributed to the methylamine-utilizing methanogen *Methanohalophilus*.

Highly saline soda lakes, which have an NaCl content ranging from 20% to saturation, have a different microflora. Cyanobacteria are absent and the dominant micro-organisms are haloalkaliphilic archaebacteria. These reach populations of 10^7–10^8 ml^{-1}. Haloalkaliphiles are classified in two genera: *Natronobacter* (**283**), which contains three species, and *Natronococcus*, which has a single species *Na. occultus*. Archaebacteria of this type occupy a highly specialized eco-niche, which does not overlap with that occupied by other halophilic archaebacteria. The ecology of haloalkaliphiles has been considered analogous to that of non-alkaliphilic halophilic archaebacteria in neutral salterns. Alkaline salterns, such as Lake Magadi (Rift Valley) and Wadi Natrum (Egypt), undergo a cycle of precipitation of both NaCl and Na_2CO_3, however, which is totally unknown in neutral salterns (Grant, 1992).

Haloalkaliphilic archaebacteria are secondary producers in highly saline soda lakes and, in the absence of cyanobacteria, halotolerant strains of *Ectothiorhodospira* are responsible for primary production. Other halotolerant, alkaliphilic eubacteria are present which, although poorly characterized, appear to represent new taxa. These organisms are present only in small numbers in lakes approaching saturation but can compete effectively with haloalkaliphiles at NaCl concentrations of 15–20% when numbers of the two types are approximately equal. Methanogenesis due to *Methanohalophilus* continues as the NaCl concentration approaches saturation and the presence of sulphate-reducing bacteria is strongly predicted but as yet unproven.

High Ca^{2+} alkaline springs

Water of this type is chemically similar to that produced during cement manufacture. Alkalinity is generated in the virtual absence of CO_3^{2-} ions which are removed from solution in the presence of excess Ca^{2+} and Mg^{2+}. In the initial stages, CO_2-charged surface water erodes surrounding minerals. The chemistry of the groundwater forming the spring is determined by low temperature, sub-surface weathering of two primary minerals, olivine and pyroxene, which are decomposed by the process of serpentization (**284**). This results in a $Ca(OH)_2$ brine in which the $Ca(OH)_2$ is in equilibrium with the solid phase. Under most circumstances, the solubility of $Ca(OH)_2$ is very low and the brine is very dilute, being effectively an oligotrophic system in comparison with the eutrophic soda lakes. The buffering capacity is very weak when separated from the solid phase $Ca(OH)_2$ and exposed to atmospheric CO_2. Magnesium ions are removed during production of serpentine, while CO_3^{2-} is rapidly removed from solution as carbonates. Conditions are predominantly reducing due to the release of Fe^{2+} ions and the hydrolysis of transient metal hydroxides with the production of H_2.

Relatively few microbiological studies have been made of high Ca^{2+} alkaline springs. Eubacteria may be isolated at levels of up to 10^4 ml^{-1}. Isolates, however, are not particularly alkaliphilic and are bacteria common in other environments. It appears likely that these bacteria are contaminants rather than members of an alkaliphilic microflora. The question of whether or not micro-organisms can grow in these environments remains unresolved.

284

$$MgFeSiO_4 + CO_2 + H_2O \rightarrow$$
olivine

$$Mg^{2+} + HCO_3^- + H_4SiO_4 + Fe^{2+} + OH^-$$

$$MgCaFeSiO_3 + CO_2 + H_2O \rightarrow$$
pyroxene

$$Mg^{2+} + Ca^{2+} + HCO_3^- + H_4SiO_4 + Fe^{2+} + OH^-$$

$$Mg^{2+} + H_4SiO_4 + H_2O \rightarrow Mg_3Si_2O_5(OH)_4 + H^+$$
serpentine

$$MgFeSiO_4 + MgCaFeSiO_3 + H_2O \rightarrow$$
olivine   pyroxene

$$Mg_3Si_2O_5(OH)_4 + Fe^{2+} + Ca^{2+} + OH^-$$
serpentine

$$Fe(OH)_2 \rightarrow Fe_2O_3 + H_2 + H_2O$$

Note: Based on Barth *et al.* (1987)

284 Geochemical evolution of high Ca^{2+} alkaline springs.

Alkaliphilic micro-organisms from non-alkaline environments

Alkaliphilic micro-organisms can be isolated fairly easily from neutral and even acid environments, especially soils, and to a lesser extent water. It is postulated that alkaliphiles are able to alter the pH value to permit growth in a very small niche. Alternatively the metabolic activity of other micro-organisms may raise the pH sufficiently for alkaliphiles to develop. The majority of alkaliphilic micro-organisms from non-alkaline environments are Gram-positive, endospore-forming rods, currently classified with *Bacillus*. Most isolates have a red or yellow pigmentation and grow over the pH range 8.0–11.5.

Non-endospore forming alkaliphiles from non-alkaline environments have been identified with a wide range of genera. These micro-organisms are present in soil and water, although at significantly lower levels than endospore-formers. In some cases the isolates are alkalitrophic rather than alkaliphilic and, in others, the term has been loosely applied. A few alkaliphilic yeasts and moulds have been isolated but the majority of non-endospore formers are bacteria. These include isolates identified with *Aeromonas*, *Corynebacterium*, *Micrococcus*, *Paracoccus*, *Pseudomonas* and *Streptomyces*.

Man-made alkaline environments are populated by alkaliphilic and alkalitrophic bacteria of the same types as are isolated from non-alkaline environments. It is assumed that these are derived by enrichment from nearby soil and water. In some cases a highly specific microflora develops. Examples are *Exiguobacterium aurantiacum* in potato processing waste and *Ancyclobacterium* in waste water from paper and board processing (Grant, 1992).

NON-ALKALINE, HIGHLY SALINE ENVIRONMENTS

Occurrence and distribution of non-alkaline, highly saline environments

Large-scale, non-alkaline, saline environments are typified by concentrated brines such as those in solar salterns and salt lakes where NaCl concentrations may be saturated. Many environments of this type are situated in tropical or subtropical regions but some are found in temperate regions; salt lakes also exist in Antarctica. Concentrated brine environments are of two basic types: thalassohaline (derived from seawater and characterized by a neutral to slightly alkaline pH value and a dominance by Na^+ and Cl^- ions) and athalassohaline (in which ions other than Na^+ and Cl^- are of major importance). Such environments include soda lakes.

Halophilic archaebacteria survive in both solar and mined salt; small-scale, highly saline environments also exist in fish salting and leather tanning operations as well as in brines used in meat curing. Hypersaline soils are found in the bottoms of dried-up salt lakes and abandoned salterns but are of relatively low NaCl content ($c.5.0\%$ to $c.11\%$). A novel terrestrial halophilic environment is represented by the plant leaf surfaces of a salt-excreting plant *Atriplex halimus*.

Halophilic micro-organisms

These may be defined as those with an NaCl optimum requirement for growth in excess of physiological levels: obligate halophiles are those unable to grow at normal physiological levels while facultative halophiles are those which are able to grow at normal physiological levels but which grow optimally at higher concentrations. Halotolerant organisms are able to grow at relatively high NaCl concentrations but grow optimally at normal physiological levels. It must be stressed that neither facultative nor obligate halophiles are necessarily able to grow in highly saline environments. Halophilic micro-organisms able to grow in highly saline environments of greater than 15–20% NaCl comprise both eubacteria and archaebacteria. Fungi and, allegedly, some yeasts are able to grow at NaCl concentrations approaching saturation but these micro-organisms are halotolerant rather than halophilic.

Although a number of halophilic eubacteria are able to grow in NaCl concentrations approaching saturation, the optimum is usually considerably lower. Some of these organisms grow over a very wide growth range and a requirement for NaCl at levels above physiological can be hard to demonstrate. Minimum NaCl concentrations are usually 2.5–4% and optimum concentrations typically 8–10% although some have an optimum as low as 5%. Bacteria of this type are commonly referred to as moderate halophiles. Some genera, such as *Halomonas*, contain no non-halophilic species, but many are members of genera common in normal environments. Gram-negative rods, such as *Vibrio* and *Pseudomonas*, are most common but a halophilic species of *Bacillus* is also recognized. A number of obligately anaerobic halophiles have been described, including *Halobacteroides* and *Sporohalobacter*. Halophilic members of specialist groups have also been isolated, including methanogens, sulphate-reducing bacteria and phototrophic purple sulphur bacteria.

Although some halophilic archaebacteria (halobacteria) have minimum NaCl concentrations for growth of $c.15$–20%, many will grow at lower concentrations than previously appreciated and, in some cases, minimum concentrations for growth are as low as 8%. These micro-organisms may be known by the term extreme halophile. The optimum is 20–25%. Four genera are currently recognized, of which *Halobacterium* is most common. All are characterized by pleomorphism and/or irregular cell shapes. In addition gas-vacuolated, square-shaped halophiles have been isolated. Cells are 3–4 µm or larger in size and have not yet been isolated in pure culture. Phages for halophilic archaebacteria have been isolated. These require high levels of Na^+ ions for stability.

Microbial communities in non-alkaline, highly saline environments

Concentrated brines (salt lakes and salterns)

Salt lakes and salterns with NaCl concentrations in excess of 20% are simple environments with short food chains. Many physiological groups, including nitrifying bacteria, sulphate-reducing bacteria and methanogens, are absent at very high NaCl concentrations and the microflora is dominated by halophilic archaebacteria. These are present in numbers up to 10^8 ml^{-1} and the water is often coloured red by the carotenoid pigments. The nature of the dominant microflora and the ecology of the system differs between thalassohaline and athalassohaline environments. There are, however, two major constraints common to each: oxygen deficiency and high light intensity. The solubility of O_2 in concentrated brine is very low at ambient temperatures, concentrations of less than 0.2 mg l^{-1} having been measured in surface waters of the Great Salt Lake. It seems likely, therefore, that halophiles are frequently exposed to stress through O_2 limitation. Many extreme halophiles are gas vacuolated which may be a factor in maintaining the micro-organisms in the upper layers of the brine, although this contention has never been proven. The same role has been proposed for motility and aerotaxis but this again remains unproven. Halophilic archaebacteria do, however, have a number of strategies for growth under conditions of O_2 limitation. Most are capable of growth using NO_3 as the terminal electron acceptor but the environmental significance seems limited since ammonia is the predominant form of nitrogen and a rapid turnover of NO_3 seems unlikely (Oren, 1994).

Many isolates are able to grow anaerobically by the reduction of dimethylsulphoxide to dimethylsulphide, fumarate to succinate and trimethylamine oxide to trimethylamine. The latter mechanism may be of importance in nature since trimethylamine oxide is likely to be present in significant quantities. Halophilic archaebacteria are also capable of fermenting a range of substrates and fermentation may well be an important means of anaerobic growth. Some strains of *Halobacterium* are able to use light energy to drive cellular metabolism when O_2 levels drop too low to support respiration. This mechanism has received considerable attention but is an uncommon feature amongst halobacteria as a whole. Photophosphorylation yields sufficient ATP to prolong viability for extended periods but cannot normally support growth. The mechanism involves synthesis of a modified cell membrane in response to O_2 limitation. This is purple in colour and is often known as the purple membrane. The visual pigment bacteriorhodopsin is responsible for coloration and for the conversion of light to electrochemical energy by pumping protons from the cell interior to the exterior. This creates a cross-membrane potential that the cell can use for other energy-requiring purposes.

The light intensity at the surface of salt lakes and salterns is very high and a general property of halobacteria is the synthesis of carotenoid pigments to act as photoprotectants. *Halobacterium halobium* contains additional retinal-based pigments such as halorhodopsin and has both blue and red light-sensitive systems. The blue-sensitive system mediates the avoidance of blue light, while the red-sensitive system mediates both an attractant response to red light and a much more sensitive avoidance of UV-B irradiation. It is 'tempting to postulate that halophilic archaebacteria use sensory rhodopsin to move into an environment of high light intensity near, but not too near, the surface where exposure to UV-B is potentially lethal. While there the cells rely on the rhodopsin pigments to run several essential ion "pumps". There is, however, little or no evidence to support this scenario.' (Oren, 1994).

Thalassohaline brines

Thalassohaline brines are derived from the evaporative concentration of seawater and all have a remarkably similar ecology. Environments of this type are typified by the solar salterns used commercially for production of salt. A number of salt lakes which are not, or are no longer, in contact with the ocean have a similar ionic composition to salterns and a similar microflora. These include the Great Salt Lake in the US and Lake Assal in Djibouti which is fed by saline salt springs.

During the evaporation of seawater in salterns, sequential precipitation of calcium carbonate and calcium sulphate (gypsum) occurs leaving a remaining hypersaline brine. Precipitation of $CaCO_3$ may involve a moderately halophilic species of *Vibrio*. Sodium chloride subsequently precipitates as halite crystals and dense brines of magnesium chloride remain. These brines – bitterns – are highly hostile to life. The microflora undergoes successional change during concentration. In the early stages the original marine bacteria are replaced by biochemically active, Gram-negative halophiles of relatively low optimum NaCl concentrations for growth. Moderately halophilic bacteria, such as *Halomonas*, increase rapidly in numbers when the NaCl concentration exceeds *c*.7%.

These organisms dominate the microflora at NaCl concentrations up to *c*.17.5%, after which extremely halophilic archaebacteria are of increasing importance. At a concentration of *c*.20% the outcome of competition between halophilic eubacteria and archaebacteria is largely dependent on ambient temperature, with low temperatures favouring eubacteria and high temperatures favouring archaebacteria. The two types are able to coexist in some circumstances. *Haloferax* (**285**) is the dominant halobacterium at NaCl concentrations of *c*.20%, but is replaced by *Haloarcula* as the salinity increases. *Halobacterium* (**286, 287**) is dominant at very high NaCl concentrations in the crystallizer ponds, although *Hx. volcanii*, which is highly tolerant of Mg^{2+}, ions may develop after the precipitation of halite. A number of halobacteria produce bacteriocins (halocins), which may enhance their competitive ability. Anaerobic eubacteria have been isolated from anoxic sediments of salt lakes and solar salterns where archaebacteria are unable to grow. Four genera are currently recognized: *Haloanaerobacter*, *Haloanaerobium*, *Halobacteroides* and *Sporohalobacter*, but others may exist.

Bacteria isolated from salt lakes and salterns are heterotrophic and primary production is largely by the eukaryotic alga *Dunaliella*. *Dunaliella* is capable of rapid growth and blooms at cell densities in excess of 10^4 ml^{-1}. The algal blooms are relatively short lived but are often followed by blooms of halophilic archaebacteria whose numbers exceed 10^8 ml^{-1}. Excretion of the osmolyte glycerol by *Dunaliella* and its rapid uptake by halobacteria is thought to be an important factor in archaebacterial blooms (Oren, 1993).

Although both amoeboid and ciliate protozoa have been reported in hypersaline brines, nothing is

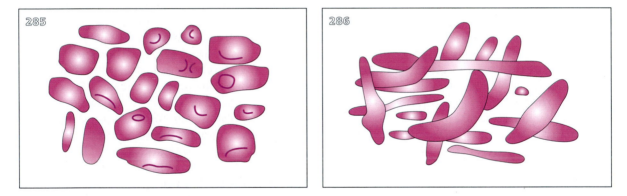

285, **286** Diagrammatic representation of the cellular morphology of extremely halophilic archaebacteria: *Haloferax* (**285**), *Halobacterium* (**286**).

287 Photomicrograph of *Halobacterium*. Like many rod-shaped halophiles, including both extreme and moderate genera, pleomorphism and the presence of irregularly shaped cells is common, especially when grown under sub-optimal conditions. Reproduced by courtesy of Prof. W.D. Grant, Department of Microbiology and Immunology, University of Leicester, UK.

known of their predatory habits (Oren, 1994). In some systems, however, *Dunaliella* is grazed by macroscopic brine shrimps (*Artemia salina*) which can reach a sufficiently high population density to justify commercial harvesting. Brine shrimps and the larvae of brine flies are an important source of organic substrates for heterotrophic micro-organisms. Considerable quantities of chitin may be deposited in sediments, and halophilic, anaerobic, chitin-utilizing bacteria (*Haloanaerobacter*) have been isolated from sediments in salterns in Chula Vista, California (Liaw and Mah, 1992). Bacterio-phages for *Halobacterium* may exist in either the virulent or temperate state, the transition between the two being mediated by NaCl concentration. Phages proliferate only at the lower end of the NaCl growth range when the halobacteria are stressed by the sub-optimal conditions prevailing. At higher NaCl concentrations the phage–host relationship is temperate. The *Halobacterium*–halophage relation-ship is thus well adapted to salterns in that produc-tion of phages is favoured in the less concentrated brine ponds, thus maximizing the chance of phage survival. At concentrations approaching saturation, conditions are optimal for growth of *Halobacterium* and the establishment of the carrier state simul-taneously protects bacteria from extensive phage-induced lysis and provides for the perpetuation of the phage (Oren, 1994). Little is known of the the role of phage in regulating *Halobacterium* populations in nature. Phage numbers increase following dilution of brine by heavy rain but the *Halobacterium* population is destroyed by further dilution. It appears, therefore, that phage develops at the expense of bacteria doomed by other factors without exerting a selective disadvantage on the host.

Stratified microbial mat communities develop in some salterns. The inability of non-specialized micro-organisms to grow at very high NaCl concentrations means that these do not develop in the most concentrated brines but in those with a NaCl concentration of c.13–20%. The community structure is generally similar to mats in environ-ments of lower salinity but selective pressures usually mean that fewer micro-organisms are involved. Hypersaline mat communities occur in a number of geographical locations including Salins-de-Giraud in the Camargue region of France, Alicante in Spain, Guerro Negro in Baja California and the Solar Lake in Sinai. The specific composition of the microflora is similar in each case, the purple layer being composed of anoxy-genic phototrophs of the families Rhodospirillaceae and Chromatiaceae. There is variation, however, in the cyanobacterial population forming the green surface layer. In the Guerro Negro ponds and Solar Lake this consists of *Microcoleus chthonoplastes* while *Spirulina* is dominant in the Alicante and *Phormidium* in the Salins-de-Giraud salterns. *Spirulina* and *Phormidium* sometimes coexist with the unicellular cyanobacterium *Aphanothece*. The cyanobacterial population is related directly to the salinity in each environment.

The Salins-de-Giraud salterns are of particular interest in that the mat develops between the sediment and a calcium sulphate (gypsum) crust, forming a physical barrier between mat and over-lying water column. This means that the mat is a closed system, no interchanges taking place with the external environment (**288**) (Caumette *et al.*, 1994).

288 Structure of microbial mat community. Salins-de-Giraud salterns, Camargue. Structure of zones:
A. *Aphanothece*; B. Gypsum crystals; C. *Phormidium*;
D. Purple bacteria (*Chromatium*; *Thiocapsa*); E. Black sediment containing FeS, gypsum and sulphate-reducing bacteria. After Caumette *et al.* (1994).

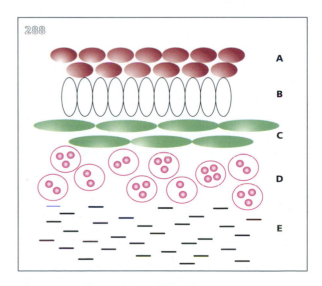

The Salins-de-Giraud mats have a population structure consisting of *Phormidium* (**289**) directly below the gypsum layer, underlain by moderately halophilic members of the Chromatiaceae, resembling *Chromatium* (**290**) and *Thiocapsa* (**291**). This layer is directly above the anoxic sediment, which contains FeS and gypsum and supports a population of the moderately halophilic, sulphate-reducing bacterium *Desulfovibrio halophilus*. Strictly anaerobic, halophilic eubacteria are also present. These organisms are fermentative and of considerable importance in supplying low molecular weight organic compounds and H_2 to sulphate-reducing bacteria. Colourless sulphur bacteria are present in significant numbers throughout the mat and the upper layer of the sediment. In summer, an additional cyanobacterial layer, consisting of yellowish-brown *Aphanothece* develops above the gypsum layer. Sulphate reduction rates in the sediment are high, sulphide produced diffusing upwards. During the day sulphide is reoxidized by oxic and anoxic oxidations, involving chemical oxidation by O_2 produced by cyanobacteria, chemolithotrophic activity of colourless sulphur bacteria and anoxygenic photosynthesis by purple sulphur bacteria. During the dark, the absence of O_2 production by cyanobacteria means that sulphide reoxidation is incomplete and sulphide migrates upward to the gypsum crust. Sulphide in the upper region of the mat is removed during the next illuminated period, it being possible that initial photosynthesis by cyanobacteria is anoxygenic, using sulphide as the electron donor. Longer term seasonal variations occur in that sulphide slowly accumulates during the winter when activity of the photosynthetic microflora is very low. Sulphide levels then fall in spring when photosynthesis is at a very high level, despite increasing rates of sulphate reduction.

Chromatium salexigens and *Tc. halophila* both contain bacteriochlorophyll *a* as photosynthetic pigments but the absorbance maxima differ. This enables the two organisms to coexist, using the same infrared light, but without competition for the same wavelengths (Caumette *et al.*, 1994).

289 The green layer consisting of *Phormidium* of the microbial mats at the Salins-de-Giraud salterns, Camargue. Reproduced with permission from Caumette *et al.*, 1994, © 1994, Federation of European Microbiology Societies.

290 Bacteria resembling *Chromatium* in the purple layer of the microbial mats at the Salins-de-Giraud salterns, Camargue. Reproduced with permission from Caumette *et al.*, 1994, © 1994, Federation of European Microbiology Societies.

291 Bacteria resembling *Thiocapsa* in the purple layer of the microbial mats at the Salins-de-Giraud salterns, Camargue. Reproduced with permission from Caumette *et al.*, 1994, © 1994, Federation of European Microbiology Societies.

Salterns and salt lakes in tropical and subtropical climates can attain temperatures as high as 41°C (106°F) and ambient temperatures are generally close to the optimum for growth of halobacteria (35–40°C (95–104°F)). Such high temperatures are the exception in temperate climates and the inhabitants of ecosystems such as the Great Salt Lake are usually growing at sub-optimum temperatures. The most striking example of low temperature growth of halophilic archaebacteria is provided by Deep Lake in Antarctica. The highest temperature attained in this lake is 11.5°C (52.7°F) and for much of the year the temperature does not exceed 1°C (33.8°F). This lake is inhabited by *Hb. lacusprofundi* which has a T_{opt} of 31–37°C (87.8–98.6°F) but which is able to grow at *c.*4°C (39°F). It has been estimated that, in its natural environment, this organism undergoes no more than eight doublings per year. Despite this very slow growth rate *Hb. lacusprofundi* is able to completely dominate the microflora due to a lack of competition from other halobacteria, which are unable to grow below 10–12°C (50–54°F).

Athalassohaline brines

Athalassohaline brines, in which ions other than Na^+ and Cl^- are predominant, vary in chemical composition. Soda lakes are a special type of highly alkaline athalassohaline brine, supporting a specialized microflora of haloalkaliphiles. *Halobacterium* and other neutrophilic halophiles are unable to grow in these environments.

The best known athalassohaline environment is the Dead Sea. This has been monitored for many years and a wealth of information is available concerning its chemistry and microbiology. The Dead Sea represents an extreme example of an environment where divalent ions contribute the majority of the cation sum and which is highly hostile to microbial life. The normal molar composition is 1.8M Mg^{2+}, 0.4M Ca^{2+}, 1.7M Na^+ and 0.14M K^+. Halophilic archaebacteria adapted to growth in thalassohaline environments are unable to grow in brine of this composition, primarily due to the toxicity of the Mg^{2+} ion. Archaebacterial inhabitants of the Dead Sea, *Hx. volcanii* and *Hb. sodomense*, are highly adapted to the environment and are not only highly tolerant of Mg^{2+} but have an obligate requirement for the ion. Despite this, the concentration of Mg^{2+} ions is the limiting factor for extensive microbial growth in the Dead Sea. This results from the inhibition of *Dunaliella* at very high Mg^{2+} concentrations and consequent loss of primary production. Development of micro-organisms in the Dead Sea is closely related to the physical structure of the water, especially stratification and the existence of an upper layer of relatively low ion concentration. During normal dry periods the Dead Sea is non-stratified and of a more or less uniform composition with a high concentration of Mg^{2+} and other ions. The water activity level under these conditions is 0.67 or lower. *Dunaliella* is not detectable and archaebacterial numbers fall to an estimated 5×10^5 to 10^6 ml^{-1}. These values are almost certainly overestimates. Significant rainfall leads to stratification with a pynocline at varying depths and a relatively dilute upper layer. This results in a *Dunaliella* bloom, which is short lived, but followed by an archaebacterial bloom when numbers reach in excess of 5×10^7 ml^{-1}. The archaebacterial bloom is longer lived, numbers falling to a stable level of *c.*5×10^6 ml^{-1} until stratification disappears. This usually involves a period of about one year. Extensive growth of micro-organisms in the Dead Sea is thus not a normal phenomenon but occurs as a result of a 'catastrophe' involving the unusual event of abundant rainfall in the catchment area (Oren, 1994).

Solar and mined salt

Sodium chloride precipitates out of solution in the crystallizer ponds of solar salterns as halite crystals. Pockets of concentrated brine are trapped during crystal growth and constitute up to 6% of the crystal weight. Halobacteria are captured in the brine pockets and remain viable for several years, crude solar salt containing 10^5–10^6 bacteria g^{-1}. Survival mechanisms in extreme halophiles have not been elucidated although it has been assumed, without any apparent justification, that parallel processes to those in non-halophiles are involved. A most intriguing development has been the isolation of live halophilic archaebacteria from rock salt buried for more than 200 million years (Norton *et al.*, 1993). Extremely halophilic archaebacteria have been isolated from the interior of intact pieces of rock salt and from sources in the mine, including salt efflorescence on tunnel walls and brine pools. A number of different halobacteria have been isolated including *Hb. saccharovorum* and species of *Haloarcula*. Some isolates are distinct from known genera. The origin of these bacteria remains obscure and it is not known whether or not they represent 'living fossils' that have survived for hundreds of millions of years (Oren, 1994).

Meat curing brines

Meat curing brines have NaCl concentrations in the range of *c*.20–27.5%. Sodium nitrite is present at levels of 0.1–0.15% and, in many cases, sodium nitrate at levels of 0.3–0.5%. The bulk of the population consists of moderately halophilic, Gram-negative rods, phenotypically resembling non-halophilic *Pseudomonas* species. The dominant microflora attains a population density of 10^8–10^9 ml^1 under normal conditions. Growth can be very rapid but is usually controlled by maintaining the brine at a temperature of less than 5°C (41°F). Growth is also restricted by the presence of sodium nitrite although the bacteria present have a high level of resistance to this salt. Brines are strongly reducing in sub-surface layers and both nitrate and nitrite are used as terminal electron acceptors. The reduction of nitrate is of technological importance in producing the active curing ingredient nitrite. Use of NO_3 and NO_2 as terminal electron acceptors is probably the most important mechanism of anaerobic growth but most of the bacteria present can also use the arginine dihydrolase pathway for this purpose.

In addition to the main population of moderate halophiles, curing brines contain a sub-population of weakly halophilic bacteria. These are predominantly species of *Vibrio*, such as *V. costicola*, which differ from marine isolates in a number of ways. Such bacteria are unable to grow in the body of the brine. The NaCl content, however, is not homogeneous and relatively low concentrations are found close to the meat where water diffuses out of the tissues.

Hypersaline soils

Hypersaline soils are restricted habitats found in the bottoms of abandoned salterns and other areas previously covered by water of high NaCl content. In many cases there is abundant plant cover provided by xerophilic plants such as *Anthrocnemum* (Quesada *et al.*, 1982). Little information is available concerning hypersaline soils but the microflora appears to consist largely of moderately halophilic and halotolerant bacteria. These have been assigned to a wide range of Gram-positive and Gram-negative genera, the most numerous of which were *Bacillus*, *Alcaligenes* and *Pseudomonas*. Many of the halophilic isolates grew over a very wide NaCl range, suggesting that spatial and temporal variation in NaCl content of the soil favours less specialized halophiles. Small numbers of extremely halophilic bacteria (**292**, **293**), identified with *Halobacterium*, were also isolated, probably from microhabitats of high NaCl concentration (Quesada *et al.*, 1982) Bacteria are present in greater numbers in rhizosphere hypersaline soil than non-root soil and it is probable that the role of bacteria in such soil, and the relationship with plants, is similar to that in non-saline soils.

Leaf surfaces of *Atriplex halimus*

The leaf surfaces of *Atriplex halimus*, a salt-excreting plant, have only recently been recognized as a terrestrial halophilic environment (Simon *et al.*, 1994). Salts and organic matter coat the leaves

292, 293 Bacterial growth on hypersaline soil. Disused salterns in Britanny, France. Although unmanaged, sea water periodically enters some of the salterns and there is considerable, visible, NaCl deposition. The red colouration results from the growth of *Halobacterium* spp., but there is also a diverse population of moderate, Gram-negative halophiles and halotolerant bacteria, predominantly *Bacillus* spp. Illustrations reproduced by courtesy of Liam Varnam.

during the dry season although overnight dew leads to diurnal wetting, followed by repeated desiccation. The leaf surface environment is therefore subject to wide variation in salinity and water activity. The microflora present is of very limited diversity being dominated by an orange-pigmented, halotolerant species of *Pseudomonas*. This has an optimum NaCl concentration for growth of 5% but will grow at concentrations up to 20% and is highly desiccation resistant. Desiccation resistance appears to be related to the protective effect of organic material from the leaf rather than being an intrinsic property of the bacterium.

REFERENCES

Barth, A.H., Cristophi, N., Neal, C. *et al.* (1987) Trace element and microbiological analysis of alkaline ground waters in Oman, Arabian Gulf: A natural analogue for cement pore waters, *Report of the Fluid Processes Research Group of the British Geological Survey*, FLPU 87–2.

Brierley, C.L., Brierley, J.A., Norris, P.R. and Kelly, D.P. (1980) Metal-tolerant micro-organisms of hot, acid environments, *Microbial Growth and Survival in Extremes of Environment* (eds. G.W. Gould and J.E.L. Corry), pp. 39–51, Academic Press, London.

Brock, T.D. (1984) *Thermus, Bergey's Manual of Systematic Bacteriology*, Vol. 1 (eds. N.R. Krieg and J.G. Holt), pp. 333–7, Williams and Wilkins, Baltimore.

Caumette, P., Matheron, R., Raymond, N. and Relexans, J.-C. (1994) Microbial mats in the hypersaline ponds of Mediterranean salterns (Salins-de-Giraud, France), *FEMS Microbiology Ecology*, **13**, 273–86.

Friedman, S.M. (1992) Thermophilic microorganisms, *Encyclopedia of Microbiology*, Vol. 4 (ed. J. Lederberg), pp. 217–29, Academic Press, New York.

Gadkari, D., Schricker, K., Acker, G. (1990) *Streptomyces thermoautotrophicus* sp. nov., a thermophilic CO- and H_2-oxidizing organism, *Applied and Environmental Microbiology*, **56**, 3727–34.

Grant, W.D. (1992) Alkaline environments, *Encyclopedia of Microbiology*, Vol. 2 (ed. J. Lederberg), pp. 73–84, Academic Press, New York.

Haber, R., Stoffen, P., Cheminee, J.C. *et al.* (1990) Hyperthermophilic archaebacteria with the under and open-sea plume of erupting MacDonald Seamount, *Nature*, **345**, 179–82.

Liaw, H.J. and Mah, R.A. (1992) Isolation and characterization of *Haloanaerobacter chitinovorans* gen. nov., sp. nov., a halophilic, anaerobic chitinolytic bacterium from a solar saltern, *Applied and Environmental Microbiology*, **58**, 260–6.

Norton, C.F., McGenity, T.J. and Grant, W.D. (1993) Archaeal halophiles (halobacteria) from two British salt mines, *Journal of General Microbiology*, **139**, 1077–81.

Oren, A. (1993) Availability, uptake and turnover of glycerol in hypersaline environments, *FEMS Micro-biology Ecology*, **12**, 15–23.

Oren, A. (1994) The ecology of extremely halophilic archaea, *FEMS Microbiology Reviews*, **13**, 415–39.

Quesada, E., Ventosa, A., Rodriguez-Valera, F. and Ramos-Cormenzana, A. (1982) Types and properties of some bacteria isolated from hypersaline soils, *Journal of Applied Bacteriology*, **53**, 155–61.

Reysenbach, A.-L. and Deming, J.W. (1991) Effects of hydrostatic pressure on growth of hyperthermophilic Archaebacteria from the Juan de Fuca ridge, *Applied and Environmental Microbiology*, **57**, 1271–4.

Simon, R.D., Abeliovich, A. and Belkin, S. (1994) A novel terrestial halophilic environment: the phylloplane of *Atriplex halimus*, a salt-excreting plant, *FEMS Microbiology Ecology*, **14**, 99–110.

Straub, W.L., Deming, J.W., Somerville, C.S. et al. (1990) Particulate DNA in smoker fluids: evidence for existence of microbial populations in hot hydrothermal systems, 1441–7.

APPENDICES

I Development of the Microflora in a Flooded Gravel Pit

The extraction of gravel, as a constructional material, is an important economic activity in some countries. Pits are drained by mechanical pumping while extraction is continuing, but flooding follows after commercial working ceases. Some are used for recreational use, such as water-skiing, or as marinas, while others have been used as land-fill sites. The gravel pit discussed in this Appendix was, after gravel extraction ceased, allowed to flood. There was also some illegal dumping of land-fill.

The initial microflora developing once the pit was fully flooded, reflected that of the surrounding soil. In this context, it should be noted that the siting of the pit caused very heavy run-off during rain. The initial microflora was dominated by Gram-negative, yellow-pigmented rod-shaped bacteria and 'coryneform' bacteria, many of which were identified as *Arthrobacter* spp. *Bacillus* spp. were also present in relatively large numbers.

As in all environments, the significance of the presence of Bacillus *must be interpreted with caution. The longevity of the endospores means that it is easy to overestimate the importance of the genus in a given ecosystem. This includes foods as well as natural environments. Equally, it can never be assumed that* Bacillus *is present only as surviving endospores.*

Over a period of six months, 'total viable counts' (based on colony counts on quarter-strength nutrient agar medium) fell by a factor of *c.* 10. This occurred over the spring and summer and could not be attributed to low water temperature. Low rainfall, however, meant that run-off from surrounding soil effectively ceased. During this period the qualitative analysis of the microflora showed increasing dominance by Gram-negative rod-shaped bacteria, especially yellow-pigmented. There was also an increase in the number of non-pigmented Gram-negative rod-shaped bacteria. All of the colonies of Gram-negative rod-shaped bacteria were oxidase-positive and presumptively identified with *Pseudomonas* spp.

Over the next six month period (autumn and winter), the 'total viable count' rose slowly to *c.* 10^8 cfu/ml^{-1}. At this stage, the composition of the microflora remained dominated by presumptive *Pseudomonas*. In subsequent months, there was no statistically significant change in 'total viable count', but the diversity of genera increased. Bacteria recovered on quarter-strength nutrient agar medium in numbers in excess of 10^6 cfu/ml^{-1} included oxidase-negative, yellow-pigmented Gram-negative rod-shaped bacteria, purple-pigmented, Gram-negative rod-shaped bacteria (presumptive *Janthinobacterium* spp.), Gram-negative curved bacteria, a number of which produced pink-, or red-pigmented colonies and pale-yellow-pigmented Gram-positive coccal-shaped bacteria, presumptively identified as *Micrococcus* spp. The final sampling, two years after flooding, indicated a stable microflora, although the Gram-positive cocci were no longer detectable.

The presence of high numbers of presumptive Micrococcus *in an aqueous environment is unusual, although the environmental importance of the genus may well be underestimated. The pattern of the development of the microflora, however, illustrates*

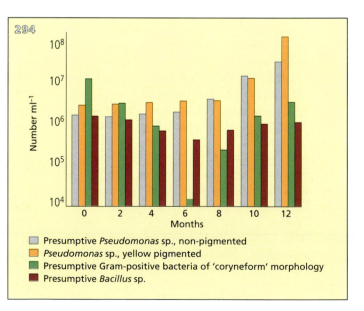

294 Development of the microflora of the water of a flooded gravel pit.

that a 'new' environment, such as the recently flooded gravel pit, can offer transitory niches for a range of micro-organisms, before a more predictable micro-flora attains full dominance.

The bar chart, **294**, illustrates the change in microbial numbers, recovered on quarter-strength nutrient agar, of the four main 'groups' of bacteria in the water of the flooded gravel pit. In particular, note the overall dominance of the presumptive *Pseudomonas* spp. And consider why there are differences in development pattern between yellow-pigmented strains (population 1) and non-pigmented strains (population 2). The bar chart **295** illustrates the diversity of the population before the disappearance of the Gram-positive coccal shaped bacteria.

295 Dominant microflora of gravel pit two years after flooding.

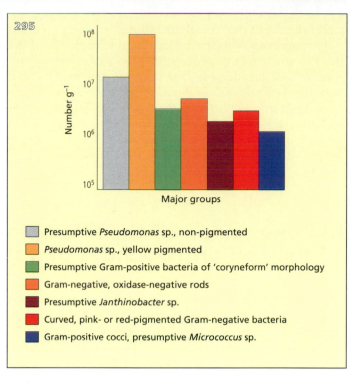

295

Presumptive *Pseudomonas* sp., non-pigmented

Pseudomonas sp., yellow pigmented

Presumptive Gram-positive bacteria of 'coryneform' morphology

Gram-negative, oxidase-negative rods

Presumptive *Janthinobacter* sp.

Curved, pink- or red-pigmented Gram-negative bacteria

Gram-positive cocci, presumptive *Micrococcus* sp.

II The Great Goldmine Mystery

As part of a survey of the microbial ecology of a mountain stream – as it made its way from its multiple sources on a mountain to its confluence with a small river in the valley below the mountain – investigation was made of the water issuing from the adit of a disused gold mine. The adit was accessible for a certain distance, the outflow appearing to originate in a closed side-tunnel.

Water samples taken from the stream above the adit outflow had a 'total viable count' (quarter-strength nutrient-agar medium) of 10^7 to 10^8 cfu/ml. This count was higher than expected, but probably reflected drainage from bog pools, the microflora of which reflected that of the underlying mud and was dominated by yellow-pigmented, Gram-negative, rod-shaped bacteria. Despite the obvious presence of peat there was no evidence of acidification.

Water from the mine adit had a 'total viable count' in the order of 10^3 cfu/ml. The microflora consisted of a mixture of white- and yellow-pigmented, Gram-negative, rod-shaped bacteria. There was no visual evidence of slime growth in the water, while the slate walls and remaining wooden structures within the adit were slime-free, although very moist.

A major surprise resulted from analysis of water immediately downstream from the adit outflow. 'Total viable counts' did not exceed 10^3 cfu/ml and there was a total disappearance of yellow-pigmented bacteria, the microflora being dominated by white-pigmented, Gram-negative, rod shaped bacteria. Enzyme profiles obtained by the API-zym™ system suggested that these bacteria shared an identity with those of the adit outflow.

Samples taken *c*. 100 m from the outflow showed an increase in number to 10^5 cfu/ml. Yellow- and white-pigmented colonies were present in approximately equal numbers. The microflora was restored to a 'total viable count' of 10^7 to 10^8 cfu/ml, predominantly yellow-pigmented, Gram-negative, rod-shaped bacteria at the next accessible sampling point; approximately 400 m downstream.

The findings were repeatable and there was no link with seasonal variations in water flow. The area is metalliferous and a number of heavy metals are present. Semi-quantitative analysis, however, showed no evidence of heavy metal pollution. Unusual features were the very sudden decline in the microbial population apparently associated with the adit outflow and the relatively rapid re-establishment of the typical microflora.

In many areas of microbiology, 'hard to explain' findings are a fact of professional life and cause considerable problems in, for example, the food industry. As is probable in this example, the answer often lies in an overall appreciation of the ecosystem, including the physico-chemical background and geobiology, rather than increasingly sophisticated microbiology. The use of genetic techniques, for

example, would undoubtedly explore the relationship between the various components of the microflora rather than use the relatively simple tests employed in this exercise. The use of genetic or other sophisticated tests, however, does not solve the central dilemma in examples such as the gold mine, where the key question may well be: 'What have we overlooked?'

III Comparison of the Microbiological Profile of the Soil on Two Sides of a Valley

The north side of a valley is forested with a mixture of coniferous and non-coniferous trees, while the south side is arable land, primarily cultivated for malting barley. Analysis was based on 'total viable counts' together with a simple classification using colony morphology and pigmentation, together with cellular morphology (**296**). The pH (10% w/v in distilled water) was also measured.

Analysis showed that 'total viable counts' were markedly higher on the north, forested side of the valley. This may be attributed to the higher levels of nutrients available and is reflected in differential counts of micro-organisms. The count of microfungi, for example, is much higher in soil from the wooded side of the valley – almost certainly due to the greater level of organic carbon in the soil and the presence

of detritus, much of which is highly recalcitrant to degradation by non-filamentous micro-organisms.

The levels of Streptomycetes demonstrate this more vividly. In the south (arable) side of the valley, numbers never exceeded 10^3 cfu g^{-1}, while in the north (wooded) side, numbers were typically in the order of 5×10^5 cfu g^{-1}. Gram-negative, yellow, rod-shaped bacteria were also markedly more prevalent in the wooded soil. In contrast, Gram-positive bacteria of 'coryneform' morpholgy were present in much greater numbers in the arable soil, as were Gram-positive, aerobic endospore-forming bacteria, presumptively identified as *Bacillus* spp. Gram-positive coccal-shaped bacteria, however, were present only in very low numbers (<10^3 cfu g^{-1}) in the arable soil, but in much higher numbers (*c.* 10^4 cfu g^{-1}) in the wooded soils of the north face of the valley. This is an unusual pattern since, during the dry years (when sampling took place) the arable soil became very desiccated, which would normally be expected to favour coccal-shaped bacteria. In contrast, the wooded soil remained relatively moist, conditions which would not be expected to favour coccal-shaped bacteria.

Environmental microbiology, like all other branches of the art and science, produces its conundrums. Another lesson not too late for the learning! Quite simply, it is often not possible to investigate all possible factors affecting microbial growth and survival. It is important to understand, however, that 'What you have got is all that you have got: but it is always more than you think'.

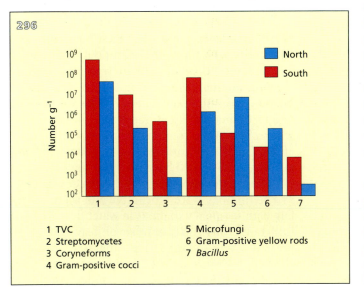

296 Microbiological profile of soil in a valley.

FURTHER READING

Textbooks

General environmental microbiology

Atlas, R.M, Bartha, R. *Microbial Ecology*, 2nd edition. Benjamin Cummings, 1987. *Very general, but a good starting point.*

Hurst, C.J. *Modelling the Environmental Fate of Microorganisms*. American Society for Microbiology, 1991. *The place to begin for would-be modellers.*

Specific principles

Brown, A.D. *Microbial Water Stress Physiology*. John Wiley, 1990. *Underpins knowledge of microbial behaviour in a number of environments.*

Bull, A.T., Slater, J.H. *Microbial Interactions and Communities*, Academic Press, 1982. *Some outdated material, but still worth consulting.*

Henis, Y. *Survival and Dormancy of Microorganisms*, John Wiley, 1987. *Of particular interest because the entire range of microbial life is discussed, not just bacteria.*

Aquatic environments

Pepper, I.L., Gerba, C.P., Brasseau, M.L. *Pollution Science*. Academic Press, 1996. *Not dedicated to microbiology, but still a very effective background text.*

Rheinheimer, G. *Microbial Ecology in a Brackish Water Environment*. Springer Verlag, 1997. *Still valuable, but beware outdated material.*

Terrestrial environments

Dommerguez, Y.R, Krupa, S.V. *Interactions Between Non-pathogenic Soil Bacteria and Plants*. Elsevier, 1978. *Fundamental discussions are still appropriate, but care must be taken over detail.*

Paul, E.A., Clark, F.E. *Soil Microbiology and Biochemistry*, Academic Press 1989. *One of the best!*

Richards, B.N. *The Microbiology of Terrestrial Ecosystems*, Longman, 1987. *Comprehensive in coverage, but limited in depth.*

Extreme environments

Da Costa, M.S., Duarte, J.C., Williams, R.A.D. *Microbiology of Extreme Environments and its Potential for Biotechnology*. Elsevier, 1989. *Based on Symposium proceedings – wide-ranging and rigorous in approach.*

Kushner, D.J. *Microbial Life in Extreme Environments*. Academic Press, 1978. *EIderly, but still a worthwhile read.*

Journals and Reviews

Applied and Environmental Microbiology. American Society for Microbiology. *Not entirely dedicated to environmental microbiology, but always a good place to start a literature search.*

FEMS Microbiology, Ecology. Federation of European Microbiology Societies. *Papers are almost invariably of a high standard, many having a strong theoretical base.*

Journal of Applied Microbiology (formerly *Journal of Applied Bacteriology*). Society for Applied Microbiology. *Publishes over a wide spectrum of applied microbiology, including high quality papers concerned with various aspects of environmental microbiology.*

Letters in Applied Microbiology. Society for Applied Microbiology. *The 'little sister' of the Journal of Applied Microbiology. Papers are similar in scope and quality, but are usually short communications.*

Environmental Microbiology. Society for Applied Microbiology. *First issue in 1999, but likely to be a valuable information source.*

Many other journals carry papers concerned with environmental microbiology. These include *Microbiology* (formerly *Journal of General Microbiology*), *Canadian Journal of Microbiology* and the *European Journal of Microbiology*.

INDEX

abyssopelagic habitat 33
Acanthamoeba polyphaga 72
Acetobacter 9
Acetogenium 134
acid rain 13, 109
acid tolerance 13, 137–8
Acidianus 133
acidification 13, 76
acidophiles 137–8, 139
Acidophilium 137, 138
Actinomadura madurae 93
Actinomycetes 24, 60, 69, 71, 92, 94
Actinoplanes 60, 61
adaptation 13, 14, 19, 20, 40
aerobes 8, 20, 23, 26
Aeromonas 62, 143
Aeromonas hydrophila 16, 54, 77
agarolytic activity of marine bacteria 33
agricultural chemicals 82, 100
agricultural runoff 67
Agrobacterium 99, 106
Agrobacterium rhizogenes 106
Agrobacterium tumefaciens 106
Agromyces rasmosus 89
Alcaligenes 59, 150
algae, cold-adapted eukaryotic 18
    colonization after wildfire 102
    extremely acidic environments 137
    metal ores 127
    photic aquatic communities 30
    SIMCOs 55
    toxigenic 32
algal bloom 31–2, 66, 74–5, 77
algicides, bacterial 30
alkaline environments, stable 140
alkaline proteases 140
alkaline springs, high calcium 142–3
alkaliphiles 140, 143
alkalitolerant micro-organisms 140
allochthonous micro-organisms 76, 89
Alteromonas 35
Alteromonas hanedai luminescence 43
aluminium, mycorrhizal fungi host plant
    protection 109
amensalism 25
Amillaria mellea 99
ammonia-oxidizing bacteria 89
amylolytic enzymes 134–5
Anabaena 36, 79, 104
Anabaena azollae symbiosis 78, 79
anabiotic state 16
anaerobes 8, 20, 23
anaerobic degradation in sediments of
    lake/pond 75
anaerobic environment, secondary
    adaptation 20
Ancalomicrobium 63
Ancylobacter aquaticus 60, 61
Ancylobacterium 143
Anguillospora longissima 65
animal:micro-organism interactions 26
antagonism 23–4
antagonistic relationships 105
Anthrocnemum 150
antibiotics 24, 77, 92, 106, 116, 134
Aphanizomenon 36, 37
Aphanothece 147, 148

Aquaspirillum arcticum 19
aquatic plants 78–9
archaebacteria 132, 133, 134, 142, 144,
    145, 146, 149
Archaeoglobus 133, 134
Archaeoglobus fulgidus 136
Archangium 98
Archezoa 20
Arcobacter 101
Arcobacter nitrofigilis 101, 104
arginine dihydrolase pathways 20
Artemia salina 147
Arthrobacter 90, 91, 100 101, 152
Arthrobacter glacialis 19
Arthrobacter globiformis 25
Arthrospira 141
arthrospores 93
Aspergillus 98, 99, 100, 138
athalassohaline brines 145, 149
athalassohaline environment 144
ATP synthesis, *Erythrobacter* 34
Atriplex halimus leaf surface 150–1
autochthonous micro-organisms 19, 76, 89
autoinducer, bacterial luminescence 44
autotrophs 8
Azolla 78, 79, 101
Azospirillum, biofertilizer use 105
Azotobacter 43, 101

Bacillus 99, 100
    brown pigment 103
    endospores 24
    flooded gravel pit 152
    halophilic 144
    hypersaline soil 150
    landfill 122
    neritic/littoral habitat 42, 45
    paddy soils 101
    soil 102, 154
Bacillus anthracis 96
Bacillus caldolyticus 132, 133
Bacillus cereus 96, 127
Bacillus coagulans 132, 133, 134
Bacillus licheniformis 97, 140
Bacillus macerans 140
Bacillus stearothermophilus 132, 133
Bacillus subtilis 97, 106–7
bacteria, abundance 68
    acid sensitivity 20
    activity in deep-sea water column 40
    aerobes 23
    allochthonous 76, 89
    ammonia-oxidizing 89
    anaerobes 20, 23
    antibiotic-resistant 77
    autochthonous 19, 76, 89
    barophilic 40
    biovolume 68
    coryneform 25, 53, 59, 90, 100, 152,
        154
    chemo-organotrophic 16
    chemolithotrophic at hydrothermal
        vents 40, 42
    DNA repair 22, 126
    endospores 22, 24, 96, 97, 102
    enteric 22, 54, 80
    epipelagic habitat 34–5

bacteria *continued*
    faecal 57, 60
    filamentous 36, 68, 69
    gas vacuoles 35, 37, 60, 61
    gliding strains 92, 94, 96
    Gram-negative 18, 61, 69, 78, 79, 99,
        115, 133, 146
    Gram-positive 18, 90, 115
    green non-sulphur 133
    halophilic 23
    hot springs 135
    iron 64
    lactic acid 26
    luminescent 43–5
    magnetotactic 72, 73
    metal ores 127
    methophilic 40, 42
    micro-aerophiles 8, 20, 23, 45, 72
    myxospores 102
    pH preference 20
    pink-pigmented 57
    plant surface colonization 112
    ponds/lakes 68, 69
    predatory 30
    protozoan, grazing of 76
    psychrophilic 18–19, 34, 40,
    purple-pigmented 57
    radiation-resistant 22
    sheathed 63, 64
    snow cover 54, 68, 69, 70
    soil 88
    yellow-pigmented 57
bacteriochlorophyll 34, 47
bacteriocins 24, 116, 146
bacteriophages, activity in aquatic
    environments 30
    algal blooms 31–2
    Halobacterium relationship 147
    nutrient availability 30
    parasitic attack 25
    sewage outlets 76–7
    soil 88
    thermophilic 132
bacteriorhodopsin 145
barley, Gram-negative bacteria 115
barophilic bacteria 40
Basidiomycetes 99
bathypelagic habitat 33
Bdellovibrio 45
Bdellovibrio bacteriovorus 26, 30, 66, 76
Beggiatoa 40, 41, 47, 104
Beijerinckia 101
benthic plant communities 42
biofertilizers 104, 105
biofilms 8–12, 102
    dehydration protection 102
    Pseudomonas in domestic water filter
        82–3
    sulphate-reducing bacteria 50, 125,
        126
bioleaching of metal ore 139
biological control 105, 106
biological oxygen demand (BOD) 80
biomarker techniques 14
bioremediation 17, 34, 100
biotechnology 17, 19, 134
birds 57, 74

black band disease of coral 51
Black Sea, anoxic sulphide-rich water 34
black smokers 40–42
Blasia 111
blooms, algal 31–2, 66, 74–5, 77
    bacterial 32, 149
    cyanobacterial blooms 31–2, 66, 75, 77
    Dunaliella 149
    phages 31–2
    seasonal 14
    soda lakes 141
    Thiocapsa 50
    white 32
Botrytis 98
Botrytis cinerea 116
Bradyrhizobium 101, 110
Brevibacterium linens 26
brine 145–9
building stone 123

Caenomorpha 125
calcite formation in freshwater 32
Calothrix 104
Campylobacter 57
Capnocytophaga 45
carbon cycle, epipelagic habitat 36
carbonates, biogenesis 32
carotenoid pigments 145
carotenoid synthesis 134
Caulobacter 45, 63
cell death in biofilms 12
cells, downsizing 24
cellular constituent assay 14
cellulase 98
Cellulomonas 90, 91, 98
cellulose 98–9
Chaetomium 98
charcoal piles 136
chasmoendoliths 124
cheese, surface ripening 26
chemo-organotrophic bacteria 16
chitin 99, 109, 147
chitinase 99
chitinolytic activity of marine bacteria 33
Chlamydomonas acidophila 137
Chlorella 26, 138
Chlorobium 34, 48
Chloroflexus 135
Chondromyces 95
Chromatium 47, 148
Chromatium salexigens 47, 148
Chromobacterium fluviatile 62
Chroococcus 141
chytrid fungi 20, 65
ciliates 68, 70
Citrobacter 13
Cladosporium herbarum 92
Clostridium, acetic acid production 134
    cellulose breakdown 102
    endospores 24
    lactic acid production 134
    mesophilic 98–9
    soil 97–8
    thermophilic 98–9
Clostridium botulinum 74
Clostridium pasteurianum 101
Clostridium perfringens type A 98
Clostridium tetani 97
Clostridium thermocellum 134
Clostridium thermohydrosulfuricum 133
cloud formation promotion 39

coal waste heaps 132, 136, 137, 139
coccolithophores 32, 39
Codium 43
cold-adaptation 18
commensalism 25, 105
competitive advantage, phosphatase
    synthesis 16
competitive strategies 23–4
compost heaps 132
copiotrophs 16
copper ore 138, 139
copper resistance 105
coral, cyanobacterial mat 51
corrosion of cast iron/mild steel 64
corrosive metabolites, solubilization 127
Corynebacterium 143
crown gall disease 106
Cryptococcus 116
cryptoendoliths 124
Cryptosporidium parvum 82–3
cyanobacteria, carbonate mineral
    formation 32
    climatic change role 39
    cold-adapted 18
    colonization after wildfire 102
    epilimnion 76
    heterocystous 36, 37
    hot springs 135
    lakes 68
    lithobiotic 122
    mat communities 48, 100, 147, 148
    nitrogen fixation 36, 43, 101, 104,
        110, 111
    photic aquatic communities 30
    seagrass 42, 43
    soda lakes 141
    sulphide-tolerant 48
    thermophilic 132–3
    tufa columns 141
    UV-B radiation 48
Cyanospira 141
Cyclidium 68
Cystobacter fuscus 95
Cytophaga 45, 59, 76, 96, 98, 99

Daphnia predation 68
Dead Sea 149
deep subsurface communities 125–6
deep-sea habitats 40–2
dehydration 102
Deinococcus 22
denitrification 39, 101–3
desiccation resistance 21–2, 102
Desulfotomaculum 136
Desulfovibrio 48
Desulfovibrio haloiphilus 148
Desulfurococcus 135
Desulphobacter 48
desulphurization of coal 139
diatoms 47
Dictyoglomus 135
Difflugia 122
dimethylsulphoxide reduction 145
Dinobryon cylindricum 70
dinoflagellates 32
diplomonads 20
dissolved humic matter (DHM) 33
dissolved organic matter (DOM) 33
diversity 14–15, 77
division, reductive 16
DNA damage, UV radiation 22

DNA hybridization 15
DNA repair 22, 126
dormancy 16
drought resistance 108
Dunaliella 146–7, 149

ectomycorrhizal symbiosis 25
Ectothiorhodospira 142
electron acceptors 51
Emiliania huxleyi 32, 39
endoliths 124
endospores, bacterial 22, 24, 96, 102
endosymbiosis 43, 126
enteric bacteria 22, 54
enteric pathogens 54, 80
Enterobacter 60
Enterobacteriaceae 54, 59, 90, 116
environmental changes, microbial
    communities 13
epilimnion 74, 76
epiliths 124
epipelagic habitat 33, 34–9
epithytic microflora 78, 79, 110, 112, 114,
    116
Erwinia 79, 115, 116
Erwinia amylovora 116
Erwinia carotovora 115, 116
Erwinia herbicola 90, 91
Erythrobacter 34
Escherichia coli 21, 54, 76, 101
estuarine environment 52
ethanol production 134
eubacteria 21, 133, 142–3, 144, 146
euendoliths 124
Euglena 20
Euglena mutabilis 137
Euglena sanguinea 22, 73
Euglenozoa 20
Euglypha 122
Eutrepia 138
eutrophication 54, 68, 77, 80
Exiguobacterium aurantiacum 143
exoenzymes, epipelagic habitat 34
exospores 22, 24, 102
extracellular polymeric substances (EPS)
    10, 12
extremely acidic environments 137, 138–9

farm animals 57
feeding strategies 30
fermentation 20, 26, 134
fermenter cultures, genetic modification 17
fern, aquatic 78, 101
ferric hydroxide tubercles 63
ferric oxides, bacterial accumulation 64–5
ferrous iron, removal by *Gallionella* 64
filter feeders 30
fireblight 116
fish, bacterial luminescence 43–5
flagellates 69, 70
Flavobacterium 58–9, 99
Flectobacillus 62, 69
flood control 67
food, drying 22
forest nutrients 109
Frankia 101, 111
freshwater, lakes 68–76
    micro-organisms and plants 78–9
    organisms 30
    pollution 76–7
    ponds 68–76

freshwater *continued*
running water 56–67
fungi, acid soils 98
cellulose degradation 98
chytrids 20, 65
cold-adapted 18
exospores 24, 92, 102
extremely acidic environments 137
lichen symbiosis 118
lignin metabolism 99
mycorrhizal 26, 108–9
plant surface colonization 112
Fusarium 98, 138

Gallionella 64–5
gas vacuoles 35, 37, 55, 61
genetic probes 15
genetic techniques 14–15
geysers 132
Giardia lamblia 82
gliding strains of bacteria 92, 94, 96
Gloeotrichia 104
glycine betaine 21, 52
glycoproteins, biofilms 10
gold mining 153–4
gold ore biobeneficiation 139
gravel pit, microflora development 152–3
Great Salt Lake (US) 145, 149
greenhouse gas, biogenesis 39
growth, energy partitioning 16
Gymnodimium catenatum 32
gypsum 32, 145, 147, 148

habituation 13
hadal habitat 33
Hafnia 60
hairy root disease 106
haloalkaliphiles 140, 142
Haloanaerobacter 146, 147
Haloanaerobium 146
Haloarcula 146, 149
Halobacterium 144, 145, 146, 147, 149, 150
Halobacterium–halophage relationship 147
Halobacteroides 144, 146
halocins 146
Haloferax 146
Halomonas 146
halophiles 25, 144
halorhodopsin 145
heathland 109
hepatitis A, E 54
heterogeneous communities 8
heterotrophs 8, 16, 34–5, 39, 76, 98
high temperature environments 132
highly alkaline environments 140–3
highly saline environments 144–51
Hirschia 80
holomictic lakes 74
hot springs 8, 132, 135, 137
humus 88, 99, 109
hurdle concept 18
hydrocarbons, long-chain 99
hydrogen, interspecies transfer 26
hydrogen ions 137
hydrogen sulphide 32, 75, 104, 136
hydrogenases 26
hydrogenosomes 126
hydrothermal environments 132, 135
hydrothermal vents 40, 41–2, 132, 134–5
hyperthermophiles 8, 133

hyperthermophilic environments 42
Hyphomicrobium 63, 80, 81
Hyphomonas 80
hyphomycetes 65
hypoferraemia of infection 16
hypolimnion 75
hypoliths 124

ice-fields 19
ice making 82
ice nucleation active (INA) bacteria 116
industry, biofilms 9, 10
inhibitors, biofilm protection 11
interception feeders 30
iron 16, 36, 39, 46, 109

Janthinobacterium 90, 91, 152–3

lactic acid production 134
lactic acid bacteria 26
Lactobacillus 26
Lactococcus 20
lakes 68–76
acidification 76
alkaline 8
clostridial toxins 74
heterotrophs 76
holomictic/meromictic 74
mountain 19
salt 145–9
soda 141–42, 149
stratification 75
water stratification 74–5
Lampropedia hyalina 72
landfill 122
lead mining, pollution 76
Legionella 71–2
Legionella pneumophila 71–2
Leptospira 22
Leptospirillum 137, 138
Leptospirillum ferrooxidans 137, 138
Leptothrix 63
Leucothrix 45
lichens, allergic responses 120
crustose 118, 119
environmental monitoring 120
foliose 118, 119
metal ores 127
parasitic 118
rainforest 118
speed of growth 120
symbiosis 26, 118
thallus structure 118
use by man 120–1
Liebig's Law of the Minimum 109
light organ 43, 44
lignin 98, 99, 109
lipids, sea surface microlayer 33
Listeria monocytogenes 9, 13, 103
lithobionts 120–2
lithobiotic habitats 122–4
littoral habitat 42–54
liverwort, *Nostoc* symbiosis 111
luminescence, bacterial 43–5
Lyngby 47
Lysobacter 30, 66–7

magnesium chloride brine 145–6
magnetotactic behaviour 72–3
maintenance, energy partitioning 16
manganic oxides, bacterial accumulation 64

marine basins, stratified 74
marine environment 33–54
marine organisms 30, 43
marine snow 40
marine waters 19
Marinomonas 35
marl sediments 32
marshgrasses 42
meat curing, brine 150
medical implants, biofilms 9
Melittangium lichenicolum 95
mesopelagic habitat 33
metabolic activity, pH changes 20
metal ions, mycorrhizal fungi host plant
protection 109
metal ores 127, 137, 138, 139
Metallogonium 138
Metallosphaera 133
methane 39, 42, 67, 75, 80, 122, 134
methanogenesis 26, 50, 72–3, 102, 122, 142
methanogens 26, 50, 67, 99, 104, 126, 142
Methanohalophilus 142
Methanothrix soehngenii 126
methophils 125
Metopus 125
Michothrix parvicella 81
micro-aerophiles 8, 20, 23, 45–6, 72
micro-algae, aerobic bacteria interaction 26
micro-organisms, morphology 14
plant aerial surfaces 112–18
microbial communities 13–15
microbial consortium 25
microbial infallibility principle 100
microbial mat community 8, 47
cyanobacterial 48, 51
diurnal cycling of oxygen and sulphur
49
neritic/littoral 42–3
oxygen concentrations 49
photosynthesis 47
phototrophs 48
salterns 147–9
sulphate reduction 48
sulphur movement 49
Micrococcus 92, 143, 152–3
Micrococcus cryophilus 18–19
Microcoleus 47, 48, 100, 147
microflora, recovery from disaster 14
microfungi exospores 102
Micromonospora 60, 61, 71, 98
Micromonospora chalcea 98
microsporidia 20
midwater habitats 39–40
mineralization 65, 80, 98
morphology micro-organisms 14
Mortierella 99
moulds 20, 65, 88, 123
mutualism 25, 105, 108
Mycobacteria 99
mycobionts 118
mycorrhizal fungi 108, 109
mycorrhizal relationships 26, 108–9
myxobacteria 66, 94, 95, 98, 102
Myxococcus 102
Myxomycetes 116, 117
myxospores 102

Nannocystis 92, 94
nanoclimate 124
nanoflagellates 30

Natronobacter 140
Natronococcus 142
Navicula 47
neritic habitat 33, 42–54
neuston 33, 73
nitrate 39, 80, 101
nitrification, midwater habitats 39
nitrifying bacteria, lithobiotic 122
nitrite 39, 82, 101
Nitrobacter 101
Nitrococcus 39, 101
nitrogen, algal blooms 31
   bioremediation 34
   cycle 36, 101–2
   deficiency in epipelagic habitat 36
   rice paddy supply 78
   turnover 101
   urea as source for phytoplankton 39
nitrogen fixation 100–1
   cyanobacteria 36, 101, 111
   neritic/littoral habitats 42–3
   rice paddies 104
   symbiosis 26, 110–11
nitrogenase 36, 42
Nitrosomonas 39, 101
Nitrosococcus 101
Nitrosolobus 39, 101
Nitrosospira 39, 101
Nitrospira 101
Nitzschia pungens 32
Nocardia 99
Nodularia 36
Nostoc 36, 111
nutrient availability 18, 30
nutrient load 54, 56, 76
nutrient recycling 34
nutrients, biofilms 11, 12
   diffusion in midwater habitats 39
   low concentrations 16–17
   soil gradients 89
nutritionally opportunist state 16

Oceanospirillum 45
oil pollution, Gulf War 100
oil wells, geothermally heated 136
oligotrophs 16
organic matter 66, 75, 80, 98
organic solutes, compatible 21
Oscillatoria 47, 48, 76
Oscillatoria limnosa 36
osmolarity 21
osmolytes 21
osmoregulation, enteric bacteria 21
osmotic pressure, definition 21
oxidation, abiotic 137
oxidative phosphorylation 20
oxidative reactions, metal ores 127
oxygen, concentration in microbial mat
   community 49
   depletion by algal blooms 32
   energy pathways 20
   gradients 8
   scavenging 20, 23
ozone layer, UV radiation 22, 48

paddy soils 78, 101, 104
pampass grass, epiphytic microflora 112, 113
paper processing waste water 143
Paracoccus 143
Paraphymonas imperforata 30
Parasponia 110

Pedomicrobium 80
Penicillium 100, 138
percolating filters, sewage 80
Perigostromidium fallax 70
pesticides 82, 100
petroleum degradation 99–100
Phaeocystis 32
phages *see* bacteriophages
pheromone, bacterial 44
Phoma betae 116
Phormidium 100, 147, 148
phosphatase synthesis 16
phosphates 13, 16, 67, 77, 80
phosphorus 31, 34, 109
Photobacterium 35, 43
photoinactivation 22
photophosphorylation 145
photoprotective pigments 22
photosensitizers 22
photosynthesis 22, 34, 36, 48, 50
photosynthetic active radiation 22
photosynthetic pigment of *Synechococcus* 34
phototrophs 48, 141
Phycomycetes, aquatic 65
physicochemical conditions 18
physicochemical gradients 8
phytoalexins 116
Phytophthora parasitica 105
phytoplankton 22, 31
pigmentation 34, 59
pigments, photoinactivation 22
pine, ectomycorrhizal associations 108
plankton, snow cover 70
Planktothrix agardhii 69
plant:micro-organism interactions 26
plant aerial surfaces, micro-organisms 112–17
plant pathogens, *Erwinia* 79, 115, 116
Plectonema 36
Pleurocapsa 141
Pleurotus ostreatus 99
poliovirus 81
pollution, atmospheric 120
   biofilm protection 11
   bioremediation 126
   faecal 57, 76
   heavy metals 9, 77
   industrial 77
   inorganic 77
   lead mining 77
   microbial consortia in sediments 77
   oil 100
   radioactive materials 77
   running water 56
   soil microflora 100
   toxic chemicals 77
Polyangium 95, 98
Polyporus adustus 99
polysaccharides 10, 135
Polystictus versicolor 99
ponds 68–76
potassium, intracellular concentration 21
potato processing waste 77, 143
predation 25–6, 30, 76
prosthecae, bacterial 62, 63, 80
Prosthecochloris 48
protein hydrolysis, ericoid mycorrhizae 109
protein synthesis in dark 135
Proteobacteria 21
Proteus 60
protozoa, drinking water 82
   endosymbiotic methanogenic bacteria 126

protozoa *continued*
   extremely acidic environments 137
   lithobiotic 122
   rumen 20
   soil 88
   sulphide-rich sands 125–6
protozoans, bacterial grazing 76
pseudomonads, fluorescent 105
Pseudomonas 20, 23, 58, 79, 90, 99, 100, 105, 116, 143
   alkaliphilic 143
   Atriplex halimus leaf surface 150–1
   bacteriocin production 116
   bioremediation for oil pollution 100
   cellulolytic 98
   chitinase production 99
   cytochrome system 23
   deleterious effects on plants 105
   domestic water filter 82, 83
   epiphytic microflora 112
   flooded gravel pit 152, 153
   hypersaline soil 150
   lignin degradation 99
   meat curing brines 150
   micro-aerophilic environment 23
   micro-algae interactions 26
   oxygen scavenging 23
   petroleum degradation 99–100
   Phytophthora parasitica suppression 105
Pseudomonas aeruginosa 58, 59, 101
Pseudomonas fluorescens 58, 116
psychrophiles 8, 18, 19, 40
psychrotrophs 18, 19, 40
Pteridomonas danica 30
purple membrane 145
pyrites 50, 104, 136, 137, 138, 139
Pyrobaculum aerophilum 135
Pyrococcus 134, 135
Pyrodictium 134
Pyrodictium occultum 133, 134

radiation-resistant bacteria 22
radioactivity 77, 126, 127
red tides 32
redox potential, lake sediments 67
reductive division 24
reed beds 23, 78, 79
rhizobacter, deleterious (DRB) 105
rhizobia, nitrogen-fixing 110
Rhizobium 101, 106, 110, 111
Rhizocarpon 118
Rhizoctonia 98
Rhizoctonia solani 107
rhizosphere 105–7
Rhodococcus 100
Rhodomicrobium 80
Rhodotorula 138
rice paddies 78, 101, 104
Rift Valley (East Africa) 141, 142
rotifera 138
Runella 69
running water 56–67

Saccharomyces cerevisiae karyotypes 23
saline lakes 8
salinity changes, neritic/littoral habitat 52–4
Salmonella 54
salt, mined 149
salt lakes 145–9
salt marshes 50
salt-excreting plant 150–1

salterns 142, 145–9, 150
Sanderia 126
Scenedesmus, *Pseudomonas* interactions 26
Sclerotium rolfsii 114
sea ice 55, 69, 70
sea surface microlayer 33–4
seamounts 132, 134–5
seawater, enteric micro-organism survival 22
seaweeds 42, 45
sediments 66, 67, 72–3, 74, 80
sediments, neritic/littoral 46–51
    anaerobic degradation 46
    anoxic 46
    electron acceptors 48, 51
    Fe–Mn zone 50
    hydrogen sulphide zones 46, 47
    iron depleted zones 47
    methanogenesis 50
    microbial mat communities 47
    mineralization 48
    oxygen zones 46, 47
    physicochemical gradients 49–50
    pyrites formation/re-oxidation 50
    suboxic zone 50
    surface 49
sequestration of metals in solution 127
Serpens 62, 72
Serratia 13, 60
Serratia marcescens 60
sewage 76
    activated sludge system 80, 81
    antibiotic resistance in bacteria 77
    biological purification of liquid 80
    chlorination of effluent 82
    disposal systems 81
    flocs 8, 80, 81
    methane production 80
    percolating filters 80
    pollution 54
    treatment 76, 80–1
    utilization of sludge 80
sheath production, bacterial 63, 64
shellfish 32, 43, 54
slime layer 59, 62
slime moulds 117
small, round, structured virus (SRSV) 54, 83
snow cover 55, 70, 71
soda lakes 141–2, 149
sodium chloride 52–4, 142, 144, 145–9,
    150–1
soft rot 115
soil 88–104, 154
    climatic change 102–4
    crumbs 8
    horizons 88
    humus content 99
    hypersaline 150
    mineralization 98
    nutrient gradients 89
    organic content 88
    permanently cold 19
    physical structure 88
    pollution 100
    waterlogging 102
solfataras 132
Sorangium 98
SOS system 22
Sphaerotilus 64
Sphaerotilus–Leptothrix group 138

Spirillum 72
Spirosoma 62, 63
Spirosoma linguale 62
Spirulina 141, 142, 147
spore formation in running water 61
Sporocytophaga 98
Sporohalobacter 144, 146
Sporosarcina ureae 96, 97
springs, alkaline 142–3
starvation 12, 16, 30
Stereum hirsutum 99
Stigmatella aurantiaca 95
Streptomyces 15, 92, 154
    antibiotic production 92
    alkaliphilic 143
    bioremediation for oil pollution 100
    cellulolytic 98
    chitinase production 99
    lignin degradation 99
    neritic/littoral habitat 45
    organic decomposition in lakes 71
    prevalence 94
    soil-dwelling 92
    spore morphology 93
    volatile compound production 92, 94
Streptomyces griseus 93
Streptomyces thermoautotrophicus 136
stress, anthropogenic 67
    competition 14, 23
    cross-tolerance 13
    diversity decrease 14
    factors 23
    littoral habitat 42
    microbial communities 13–14
    oxidative 22
    starvation 16
stress proteins, low/high temperature 19
Sulfolobus 133, 135
sulphate reduction 40, 73–4
sulphate-reducing bacteria 43, 48, 50–1,
    67, 99, 125, 136, 148
sulphide, hypersaline mat communities 147
sulphur 39, 40, 49–50, 132, 135
sulphur bacteria 34, 47–50, 73–4, 148
sulphur dioxide 13, 109, 121
sunscreen compounds 22
superthermopiles 133
survival mechanisms 19
symbiosis 25, 26, 42, 78, 79, 101,
    110–11
Synechococcus 32, 34, 135
synergism, interspecies hydrogen transfer 26

taxonomy, phenotypic 14–15
temperature 18–19
terregens factor 25
thalassohaline brines 145–9
thalassohaline environment 144
thermoacidophiles 14, 24, 133, 137
Thermoactinomyces 100, 122, 134
Thermoanaerobacter 134
Thermoanaerobacter ethanolicus 133
thermocline 33, 39, 74
Thermococcus 134
Thermodesulfurobacterium 136
thermophiles 116, 122, 132, 133, 134, 137
thermophilic environments 135
Thermoplasma 131, 136, 137
Thermoproteus group 135

Thermotoga maritima 133
Thermus 133
thiobacilli, lithobiotic 122
Thiobacillus 48, 137, 138
Thiobacillus ferrooxidans 137, 138, 139
Thiocapsa 47, 50, 148
Thiocystis 47
Thiomicrospira 47
Thiothrix 40
Thiovulum 40
toluene degradation 126
toxic chemicals, pollution 77
toxin production by blooms 32
transgenic plants 106
Trichoderma harzianum 114
Trichodesmium 36, 37, 38, 39
trichomonads 20
Trichosporon pullulans 19
trimethylamine 145
triterpenoids 116
Trypanosoma 20
tuber rot of potato 116
tufa columns 141

ultra-microbacteria 16, 17
uranium ore 138, 139
urea 39, 96, 101
Usnea distribution 120
usnic acid 121
UV radiation 22, 34, 48
UV-B irradiation 22, 48, 145

vegetation breakdown 116–17
vents *see* hydrothermal vents
vesicular–arbuscular (VA) mycorrhizae 108
Vibrio 17, 35, 43, 52, 144, 145, 150
Vibrio cholerae 43, 54
Vibrio fischeri 43, 44
Vibrio harveyi 43–4
Vibrio parahaemolyticus 43
Vibrio psychroerythrus 19
Vibrio vulnificus 17, 43, 54
viruses, aquatic 30, 31
volcanoes, underwater 132

water, availability 21–2
    channels, biofilms 10
    drinking 82–3
    urban waste 77
water activity a_w 21
water moulds 65
waterlogging, soil 102–4
white smokers 40–1
wild animals, pathogen source 57
wildfire 13, 102

Xanthomonas 115
Xanthomonas campestris 115
xylans 99, 135
o-xylene degradation 126

yeasts 19, 20, 88, 112, 113, 114, 116,
    138
Yersinia enterocolitica 60
Yersinia intermedia 60

zymogenous flora 19